The Black Book
of
Reliability Management

The Little Black Book
of
Reliability Management

(What Do You Have a Right to Expect?)

Daniel T. Daley, PE, CMRP

INDUSTRIAL PRESS

Library of Congress Cataloging-in-Publication Data

Daley, Daniel T.
  The little black book of reliability management / Daniel T. Daley.
      p. cm.
   Includes bibliographical references.
   ISBN-13: 978-0-8311-3356-6 (softcover) 1. Reliability
(Engineering)--Management. I. Title.
   TA169.D357 2007
   620'.00452--dc22

                                                              2007043053

Industrial Press, Inc.

989 Avenue of the Americas

New York, NY 10018

Sponsoring Editor: John Carleo

Copyeditor: Bob Green

Interior Text and Cover Design: Janet Romano

Dedication

To my granddaughter, Lauren and all the dedicated people who help her and other visually-impaired children.

Contents

INTRODUCTION

Occasionally, I find myself referring back to one or another of the "text" quality books on reliability. There are several excellent books of this kind in the market place and they provide a comprehensive treatment of many of the techniques used to analyze and manage reliability. These books are steeped in statistics and the rigorous mathematics needed to perform a reliability analysis. As the science of reliability has gained a foothold among all the other engineering disciplines being taught to young people, these books have become the texts used to guide the course work. As a result, they are as important as textbooks used in courses like statics, dynamics, thermodynamics, and fluids.

Definitions:
Throughout this book, I will insert "definitions" at points where I believe they will assist the reader. These are not intended to be dictionary definitions. They are intended to describe how the term is being used in the context of the current discussion.

But recalling back, when I began working as a mechanical engineer, the things I learned in college and the books I used were of limited value. There was a group of older, experienced engineers, who taught me how to apply the subjects I learned in college to real world applications.

When I was a young Air Force officer I designed a number of fairly complicated air conditioning systems for the large computer buildings housing old mainframe computers of the distant past. Although my course work in thermodynamics helped me understand how the physical world worked, it did not show me how to perform building heat load calculations, or how to apply off-the-shelf air conditioning equipment. The "old heads" were good enough to help me with that.

> **Mechanical Equipment Engineer:**
> This is a title applied most often to mechanical engineers with expertise and accountability for rotating equipment (pumps, compressors, turbines). In some situations, the title and responsibility is expanded to cover pressure retaining equipment (piping, pressure vessels and boilers).

Now the subject of reliability engineering is passing into a phase where young engineers are being employed as "Reliability Engineers" at a variety of companies. The older engineers have spent most of their careers as "Mechanical Equipment Engineers" or "Heat Transfer Engineers" or some other title that tended to focus expertise into an area associated with a specific kind of equipment or technical discipline. These individuals were concerned with reliability but, they were seldom familiar with all the techniques and tools described in the reliability texts mentioned above. As a result, the "old heads" are seldom capable of helping the new "Reliability Engineers" use all the skills and techniques they bring with them from their university learning. Relatively few companies or facilities have comprehensive reliability programs in place that apply all the talents available.

> **Reliability Engineer:**
> In the context of this book, a Reliability Engineer is one who applies specific analytical techniques like Reliability-Centered Maintenance (RCM), Reliability Block Diagram technique (RBD), Weibull Analysis and others, to all kinds of equipment to identify defects, optimize lifecycle, and create proactive maintenance programs.

Viewing this issue in another way, if a group of people working at a facility, who were not familiar with the application of engineering disciplines, went out and purchased copies of all the engineering texts available and read them cover-to-cover, little, if any engineering would be applied. Utilization of engineering principles is one-half understanding the principle and one-half knowing how it fits in the application. The same is true with the relatively-new discipline of reliability. Without knowing how to apply the science and its techniques, it is unlikely that much benefit will be derived. It would be like someone reading how to dance out of a book, never seeing someone dance in person, but then trying to dance. It would not look quite right.

This book is intended to bridge the gap between a working environment, void of the discipline needed to benefit from reliability methodology, and the detail and discipline inherent in rigorous textbooks.

Although this book suffers from the same limitations as any other book on reliability (it is like telling you how to dance rather than showing you), my objective is to describe rhythm, not the foot positions. I hope to briefly describe what it takes to get a reliability program started (collecting, cataloging, and using the data), not how to become a statistician or an analyst.

CHAPTER 1
WHAT DO YOU HAVE A RIGHT TO EXPECT?

Expect what you inspect.
An old inspector's saying

Approximately twenty years ago, the company where I was employed sponsored a workshop focusing exclusively on reliability. When the workshop and its subject were announced, I was a little surprised. There were already a large number of engineers actively engaged in "reliability" related activities. I thought that most people already under-stood and supported "reliability" as a core value. I wondered why a "Reliability Workshop" was needed.

Now, twenty years later, I have a much better answer to that question. Many if not most companies still need to have a reliability workshop, and the objectives of each workshop should attempt to answer the question, "What level of relia-bility do I have a right to expect?" To answer that question, companies must first understand all the elements that affect "reliability" and then evaluate how well they are dealing with each of them.

What do you have a right to expect?

"What you have a right to expect" is the result of the relia-bility characteristics that were designed and built-in to your systems and equipment, and have been preserved through proper operation and maintenance. If you are able to recognize the things that cause failures, you will be able to make a realistic evaluation of what you have a "right to expect" because you can determine the amount of failure prevention that has already been applied. You will also have a head start on achieving the level of performance you desire.

When a device is designed and manufactured, the included components have a certain level of robustness and the configuration provides a certain amount of redundancy. The combination of component choices and configuration leads to the characteristic best described as "inherent reliability". No matter how well you operate and maintain a system, it cannot perform better than its inherent reliability. If the system is operated and maintained as well as possible, you will harvest the maximum inherent reliability. If you operate or maintain the system in a manner less than optimum, you will experience a reliability performance that is something less than the inherent reliability.

Each part of an item of equipment tends to deteriorate over time. There are things you can do to minimize the deterioration. Generally, this deterioration will lead to failure at some point in time. If you understand the deterioration rate and the current status of deteriorating components, it is possible to intervene before the failure takes place. These actions that minimize the deterioration, or intervene before failure, are best described as "proactive maintenance". Proactive maintenance that is intended to simply monitor the current situation is a form of "predictive maintenance". Proactive maintenance that is intended to change deteriorated components before they fail are forms of "preventive maintenance". To capture all the available "inherent reliability", you need to implement an optimum program of proactive maintenance.

Inherent Reliability:
The inherent reliability of a system or device is a product of its configuration, construction, and component selection. If a system is well-designed, well-made, and constructed of quality materials and components, it is more likely to have a high inherent reliability than one lacking in any of those areas. If the inherent reliability is poor, no amount of maintenance, inspection, or excellent operation, will result in performance greater than the inherent reliability. The inherent reliability defines the upper limit that a device can achieve without a physical change.

Despite how "perfect" you believe your proactive maintenance program to be, there are always new defects or forms of deterioration that you did not expect. These new defects will result in unexpected failures, and the way you deal with them through "reactive maintenance" will affect the reliability. If your reactive maintenance system responds properly, does a good job of diagnosing and troubleshooting the problem, repairs it correctly, verifies the repair, records "as found" and "as left" conditions (so that the deterioration rate can be calculated), keeps good records, and installs proactive tasks that will intervene before the next failure, then you will harvest all the available inherent reliability.

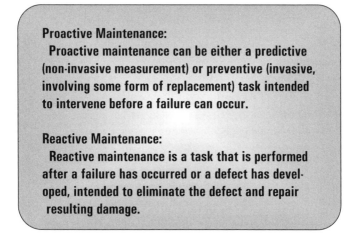

Proactive Maintenance:
  Proactive maintenance can be either a predictive (non-invasive measurement) or preventive (invasive, involving some form of replacement) task intended to intervene before a failure can occur.

Reactive Maintenance:
  Reactive maintenance is a task that is performed after a failure has occurred or a defect has developed, intended to eliminate the defect and repair resulting damage.

Saying this statement much more succinctly, if you have calculated the expected inherent reliability and you are certain you have good proactive and reactive maintenance programs, then you know "what you have a right to expect" for reliability performance.

"Reliability" or "reliability"

Many people who are not reliability experts tend to group several other characteristics under the heading of reliability. For the sake of discussion, I use the term "Reliability" (with a capital R) to describe the concept of reliability that includes all these elements.

First, reliability (with a small r) is defined as a measure of the instantaneous likelihood that a system or device will fail in a given period of time. My best analogy for reliability is a die (half of a pair of dice). Assume that the number one is a defect and when the one comes out on top a failure will occur. The reliability then is five-sixths and the unreliability is one-sixth. As long as the defect exists, there is some likelihood that a failure will occur. The only way to eliminate or reduce the likelihood of failure is to eliminate the defect.

"R"eliability:
  The termed used in this discussion to represent the combined characteristics of reliability, availability, and maintainability.

"r"eliability:
  The term used in this discussion to represent only the characteristics specifically associated with the instantaneous likelihood of failure.

Second, many people tend to roll the characteristic of "availability" into their perception of a system's Reliability. Availability is a ratio of "up-time", or time the system or device can perform its intended function, to "total time". Total "down-time" or out-of-service time is the sum of all planned down-time and unplanned down-time. "Planned availability" is determined by how long a system can operate between planned outages, and how much time is needed to conduct the outage. "Unplanned availability" is determined by how frequently unplanned outages occur (reliability), and the amount of time needed to respond to unplanned interruptions.

Third, many people also tend to roll many of the characteristics of "maintainability" into their concept of Reliability. Maintainability is a measure of your capability to return a system or device to full inherent reliability in a ratable period of time. If you were to say, "It will take three hours to fix it, but I don't know how reliable it will be", the device would not be maintainable. Also if you were to say, "I don't know how long it will take to fix it, but when I finish it will be right", it is also not maintainable. To be "maintainable", you need to be able to both restore the inherent reliability and to do it in a known amount of time.

All the elements of Reliability are not strictly parts of reliability, but many asset managers tend to include those characteristics when demanding Reliability improvements. If you are attempting to answer the question, "what do you have a right to expect?" and the question is intended to address aspects of availability and maintainability, you will need to be ready to answer some additional questions.

Drinking out of a Firehose

You are probably asking yourself, "How can this individual expect to explain the whole subject of reliability in such a small text?" The answer is that I am not. I have two objectives for this text. The first is to describe a good starting point for individuals and organizations who are willing to admit they are still new to reliability. Second, I am going to try to fill a gap that most reliability experts have ignored. That gap is the one between being a highly reactive organization (and gathering little information) and having sufficient information to begin the journey to becoming a proactive organization. I will describe an approach that will be useful and valuable to reliability engineers, as well as being worthy of resource investment by asset managers.

A number of years ago I purchased a text entitled "The Little Black Book of Project Management". At the time it was published and for sometime thereafter, there were few comparable texts on Project Management. Few individuals seemed to come "equipped" with project management skills. As a result, I repeatedly loaned the book to subordinates to help increase their knowledge and improve their skills. As usual, that practice turned out to be a good way to lose a book, so I was without a copy for several years. A few years ago I spotted another copy on the shelf of a used bookstore and purchased it. Again, I am regularly making the mistake of lending my copy out. In exchange, I get improved performance by young engineers assigned to manage projects.

My objective here is to create a text that is as useful to others who are trying to improve reliability performance as the "Little Black Book on Project Management" has been for me. The key characteristics are:
- The book is relatively short.
- Application of the approach described in the book is straightforward.

- Usefulness is not limited by scale. The book is equally useful to both large and small enterprises.
- You don't need to be an expert to use the knowledge

Prepare for Change

Like most things of any value, application of the techniques found in this book require change. The most significant change confronting the individual is a change in roles. The most significant change confronting an organization is a change in the corporate culture. To implement the approaches described in this text, both kinds of change will be required. A number of individuals will need to add tasks and change the way they are performing some of their current tasks. The organization will need to become unwilling to accept sloppiness in gathering facts and using information.

It is important to keep in mind that the tasks here discussed are closely integrated with tasks currently being accomplished within most organizations. Current meetings, current planning and scheduling protocols, and current organizational structures, will need to be modified in a thoughtful, integrated manner if optimum results are to be achieved.

One last issue …. when you think about how things have changed in the last fifteen or twenty years, there are few that have changed as dramatically as those that have been affected by computerization. Before Reliability Centered Maintenance (RCM) became popular, there were a few organizations that seemed to be light-years ahead of everyone else in terms of equipment reliability. One might ask how they achieved their performance. The answer was that those companies exercised the patience and discipline to track failures and their causes. At some point they were able to recognize patterns and relationships that were hidden in the data, and use that information to prevent failures before they occurred.

One cannot over-emphasize the dedication needed to record, store and analyze information before computers became available. There were cabinets full of paper files. The files in those drawers were faithfully maintained and the data was transferred to manual graphs that made the patterns and relationships more apparent. Most organizations that successfully accomplished this effort were led by a single-minded individual and staffed over a long period by a group of highly-dedicated people.

In today's work environment, individuals are seldom allowed the luxury of single-mindedness. They are expected to dilute their thoughts and standards to fit in with other members of their team. There are far fewer individuals in each work group, and few assignments last more than a few years. So the "corporate memory" must come from some other mechanism.

Fortunately, many of these shortcomings can be addressed by supporting the processes with computerized files. In fact, without the use of a computerized file system, this initiative would be a foolish undertaking. Few organizations have the determination and discipline to make it work without having key functions automated by a computerized filing system.

This is not to say that a well-designed computerized system will eliminate the need for human interaction and administration. There are a wide variety of elements that can go astray if they are not properly managed.

One example we will discuss is "bucketing", a term used for classifying initial failure reports (Failure Notifications) and closing reports (Failure Modes). One way in which many systems become corrupted is by allowing too many people to define classes of failure. If individuals are allowed to create a new class every time they cannot find an exact fit, there will soon be too many classes that are only slightly different from each other. When only a portion of each true failure mode is assigned to each of a number of similar failure descriptions, the final statistics can point to an incorrect failure description as being the most statistically likely. This improper result will cause inappropriate corrective actions to be taken.

The most successful system will combine a well-designed computerized database and the right amount of human interaction to ensure it is not misused or corrupted by individuals who lack an overall understanding of the system design and objectives.

If this is beginning to sound like a lot of book-keeping, it is. Good reliability management is a matter of understanding how your equipment fails. This understanding must depend on facts, not speculations or beliefs. In many ways, your equipment, and the way it operates and is maintained, is unique. As a result your reliability information will be unique.

CHAPTER 2
PATTERNS AND RELATIONSHIPS

There is nothing new under the sun.
Ecclesiastes 1:9

The usefulness in this knowledge comes
from recognizing repetitive patterns and their
relationships with significant events.
DTD

Patterns in the Stars

Several years ago, my wife gave me a telescope for Christmas. That same year, my daughter gave me a book on astronomy. Being a mechanical engineer, I had never spent much time thinking about astronomy before receiving this book. The book described the shapes of the various constellations and how to find them in the night sky. The book also described how the constellations were useful in keeping time and telling seasons from their movements and positions. It also explained why various star groupings were associated with more familiar objects like animals, tools, and weapons. The myriad of stars would remain totally random and meaningless had not the shapes of familiar objects been used to capture the outlines of the star groupings. The familiar shapes of constellations allowed the ancients to watch their movements and positions.

To me, these thoughts were a revelation. It was the first time I had ever thought of a star-filled night sky as a field containing millions of pieces of data. That data only became useful when specific patterns were identified, and then only when those patterns were found to have a relationship with some profound event, like the passage of the seasons.

Patterns in Behaviors Leading to Failures

This talk about stars and seasons is interesting but how does it apply to reliability? If one were to look at all the data associated with mechanical systems (pressure, temperature, vibration, oil condition, faults, status changes, etc.) the points of data might appear as random as the stars in the night sky. This data would begin to make sense if it was viewed in the context of an established pattern (like a constellation). It might even be possible to associate some of the patterns with significant events like breakdowns or failures.

Extending the night sky analogy, one might conclude that there are literally millions of patterns and it would be impossible to map them all. That conclusion was probably also true for the first few million times that a human being looked at the night sky. The viewers were probably overwhelmed with the sheer number of stars, let alone the complexity of their patterns and relationships. But in reality, there is nothing new under the sun … or the night sky. All the patterns and relationships are repeated over and over again every evening and every season. All it takes is the patience and discipline to map the patterns and to monitor how they behave.

In applying this logic to reliability, one failure may seem to be distinct from all other failures but in fact it is just one of many failures following a similar pattern. As with stars, there is no need to map all the pieces of data and all the relationships. It is necessary to map only the ones that are related to significant events like failures. And even then, it is unnecessary to plot all points of data, just enough to sketch out a pattern and clearly identify its relationship with a significant event.

For example, numbers of stars surround the North Star. In finding the direction north, the information provided by the positions of these other stars is unnecessary. Similarly, when vibrations reach a certain level on a pump bearing housing, fluid temperature and pressure information is not needed to point to an impending failure. One critical piece of data is sufficient. In the navigation business, a single absolute direction finder like the North Star might be good. In the reliability business, the single absolute indicator is more frequently a bad thing. By the time this clue becomes apparent, it may be too late to take actions that will prevent a failure. To prevent failure and associated damage, it is better to look for more subtle clues that become apparent farther ahead of a failure.

Unlike early man, we have the power of the computer at our disposal to assist in identifying patterns and tracking them to their ultimate disposition. We also have data mining tools that are helpful in extracting consistent patterns and relationships from tons of apparently unrelated data. Our challenge is to collect the data in a form, and with sufficient rigor and discipline, to ensure that the data is useful after it has been collected.

When preparing to collect and file data, it is important to keep the definition of reliability in mind. "Reliability" is the statistical likelihood that a system or device will complete its intended mission over a specific period of time. Again, recalling the analogy of a single die (one half of a pair of dice), if a failure-causing defect is represented by one of the numbers, say the one, a failure will occur whenever the one appears on top after a roll. In this event, the reliability would be five-sixths or around 83%. Saying it another way, the unreliability is one-sixth or a little less than 17%.

An important point to take from this discussion is that "reliability" is purely a statistical measure. To be of value, there must be a statistically meaningful amount of data and the data has to be gathered correctly. To be useful, the data must be in the proper *form* and sufficient *volume*.

Format of Data

The first characteristic we will discuss is form or format of the data. We need to collect data in a manner that allows us to distinguish the frequency of various behaviors and patterns of failure. We need an understanding of the patterns that are possible before we begin collecting data.

Maybe an example will help to better describe this issue. Let's say the early astronomers simply said that Libra was a constellation consisting of four stars, without specifying the shape of a scale, then any four-star group would fit the definition. Instead, the shape of the scale was used to provide viewers with a clearer picture of the specific shape they were seeking.

Extending this discussion to a reliability example, one might want to say that one category of pump failure is a situation where the pump stops turning. and another is when the pump can turn but does not provide adequate discharge pressure. In one instance, the anticipated behavior includes a locked shaft. In the other, the shaft is not locked, but the pump is still unable to perform its required function. Both are realistically possible behaviors, but each distinct behavior would lead to a significantly different failure mode and repair procedure.

Once the realistically possible paths from malfunction behavior to failure mode have been defined, it is possible to collect and separate data into the various distinct paths. Then we can calculate how many times the failure follows one path as compared with all other paths. Knowing the various paths, and the relative frequency of their occurrence, will help us diagnose the most likely problem and recommend the corrective action that is most likely to solve the problem in the specific instance.

Quantity of Data

The second characteristic of data mentioned above was volume or quantity of data, where the amount will depend upon the situation. If there is only one way that something can fail, you can map the failure path by recording one event. If there are two possible failure paths, more examples will be required to identify which is most likely. Let's assume one failure path is A and another is B. Also assume that the first failure path recorded was A and the second was B. The third can be either A or B so we need still more information. Assume it is A. The fourth can be either A or B again so they are equally likely. If four out of the first five was A, it would be increasingly likely that the most frequent failure path is A. As the number of failures mounts and the two 'buckets' are filled, you will become more confident which path is most likely.

As additional failure paths are added, the amount of data needed will increase. It is important to be certain that the failure paths selected are truly distinct and not just imperfect duplicates of other paths. If two paths are almost the same and would lead to the same diagnosis, troubleshooting, repair,  cause, and root cause, they should be counted in the same category. After all, the objective is to understand which response is most likely to be successful.

Independent of the kind of business you are in, repeated equipment failures give the feeling of deja vu. Unfortunately the more reactive you are, the more frequently you experience that feeling. In a highly reactive environment, you experience the feeling that bad things are happening over and over again because they are. Being reactive rather than proactive means that you do not learn from bad experiences.

The "Failure-Map" based Reliability Process

To begin the process of mapping the failures that currently affect your operation, assume that you are already working in a very proactive environment. In this setting, most of the failure patterns have already been mapped and you have identified proactive maintenance that will prevent most of the unexpected failures. Because you are aware of the required maintenance work well in advance, all the proactive work can be well planned and tightly scheduled. This condition results in highly effective and efficient maintenance.

In your world, only five percent or so of all work is reactive and unexpected. As a result, you are dealing with only one or two unexpected failures each day. So, how would you respond when those failures occur?

The overall response would begin by having the person with most knowledge of the failure file a "malfunction report". (A more detailed description of these elements will be provided in later chapters.) As with a specific starting point on a map, the malfunction report needs to be structured in such a manner that it clearly specifies the starting point for the corrective action.

 A malfunction report may be thought of as being analogous to one of several direct air routes leaving a specific city, and there are only so many places you can reach from a specific starting point. For example, if I am flying from Omaha, there are direct routes only to Dallas, Chicago, Minneapolis, St. Louis, and a few other smaller cities. Some of the routes carry more passengers and some of the routes have fewer passengers. (Similarly, some of the paths from a malfunction report occur more frequently and some less frequently.) The main issue to keep in mind is that it is possible to map a typical path for each and every destination. Once you determine that a passenger is leaving from Omaha, based on the routes of planes leaving there and based on the ridership on those routes, it would be possible to make a statistical prediction of the passenger's destination. Similarly, once you know the failure symptoms (affected function and specific behavior) in detail, and have mapped all the possible failure paths that begin with those symptoms, it is possible to identify the likely paths that will result.

The next step in your process would be to have a diagnostician review the malfunction report and the database that provides a map of all possible paths from that starting point. In a highly-mature and well-mapped system, the diagnostician will do one or more of the following three things:

1.   If there is a way to correct the malfunction immediately, the diagnostician will direct immediate actions to be taken. For instance, there

are increasing numbers of situations in computer-controlled systems where a malfunction can be corrected by simply "recycling" the computer (turning it off then restarting it). Although this action does not remove the inherent defect in the software, it does restore functionality and prevents further loss on this one occasion.

2.   If there are alternative ways to repair the malfunction, the diagnostician would select the one that is most appropriate. In some situations there are "fast track" or "do it now" crews that are capable of performing small repairs in a short time. In this situation, the diagnostician would recommend the repair approach that is able to make the needed repair in the smallest amount of time.

3.   If there is more than one path to repair, some troubleshooting will be needed. Troubleshooting is typically an invasive technique requiring time for disassembly and investigation, so the troubleshooter should start on the most likely failure mode first. Knowing both the number of paths and the relative likelihood of each path, the diagnostician will recommend the sequence of troubleshooting based on the statistical likelihood of the various paths that are possible.

In each of the steps described above, the diagnostician is like a football coach calling plays. The best play is based on the current position on the field (the symptoms described in the malfunction report) and the relative suc-

cess of the various appropriate plays in the play book (the likelihood of all possible paths from this starting point).

The next step in the process is for the troubleshooter to work through the possible alternatives. Depending on how many paths exist, it is possible that he or she might run up several blind alleys before finding the specific defect. When the troubleshooter identifies the defect, he or she will do two things:

1. Record the nature of the defect or the "failure mode" (Component-Condition).
2a. Write a work order task for a mechanic to make the repair or,
2b. If the repair is small, the troubleshooter should make the repair right then and there.

Diagnostician:
 In this context, a diagnostician is a person who analyzes external data, like historical records, remote downloads, records of parts usage, and descriptions of symptoms from operators, to identify the most likely problems and the sequence in which troubleshooting should be accomplished to minimize the downtime.

Troubleshooter:
 In this context, the troubleshooter is a person who receives direction from the diagnostician and begins the search for the defect by physical contact with the equipment and invasive steps. The troubleshooting should begin with the most likely (or frequent) path for the reported symptoms.

By this time, we will have described the portion of the defect identification and repair path that is most apparent. By doing the steps described above in a highly structured and highly disciplined manner, it is possible to completely map all failures and to become very proficient at diagnostics, troubleshooting, and performance of reactive maintenance. Most folks don't want to stop there; they want to force the transition from reactive maintenance to proactive maintenance. To make this change, they need to know how failures are occurring and what is causing them, and two further steps are then required.

The first additional step is failure analysis, which will determine the failure mechanism. The failure mechanism is the process by which deterioration is taking place. For instance one failure mechanism for a mechanical device is corrosion. If the configuration of a mechanical device provides a cathode, an anode, and an electrolyte, a natural "battery" will be formed and corrosion will occur.

Failure Analysis:
   In this context, Failure Analysis is the act of identifying the Failure Mechanism. The Failure Mechanism is a physical process (like corrosion) that causes the deterioration leading to the defect that ultimately results in the failure.

The second additional step is root cause analysis. In this step, we will determine "why" the failure mechanism was allowed to occur, or why deterioration was allowed to progress to failure.

Root Cause Analysis:
   Root Cause Analysis (RCA) is the search for the action or absence of action that created a defect or allowed deterioration to proceed to the point that a failure- causing defect exists.

This point will be discussed later, but very briefly, there are always three levels of root cause:

1. Physical Cause – Physical systems cannot defend themselves so physical components are always first to be "blamed".

2. Human Cause – No physical system ever designed itself, operated itself, maintained itself, or inspected itself, so a human is always involved in every failure.

3. Latent or Systemic Cause – Very seldom do humans intentionally do things that result in failures, so there is always a cause for failure in the systems in which they work. It could be lack of training or too little time. Conscious choices might have been made to take shortcuts. If systemic causes are not addressed, failures will continue, and the same systemic causes will result in a variety of failures.

Following all these steps not only collects the information needed to map all failures, it provides a system to handle things in the most expeditious manner. In addition, identification of failure mechanisms and root causes will cause defects to be eliminated and proactive tasks to be identified.

At the start of this discussion I suggested, "For a moment, let's assume that you are working in a very proactive environment". Now for a moment, let's assume that you are not working in a very proactive environment. Let's assume that you are working in a very reactive environment. Ask yourself, "How do I map the various paths from Malfunction Report to associated Failure Mode?" and

"How do I accurately identify the failures falling into each category without acting as described above?"

That was a rhetorical question. The answer is that you can not. Unless you begin acting as if you are proactive, you will never become proactive. Unless you slow down the repair cycle and perform repairs in a manner that restores the inherent reliability, you will never get out of the "death spiral" that ultimately leads to performance becoming increasingly worse and worse. Unless you begin taking the time to record "as-found" and "as-left" conditions, and to note the actual failure mode and failure mechanisms, you will never understand how and why things are failing.

You can choose to start acting in a proactive manner today or you can begin tomorrow. Or you can choose to begin a year from now. The journey from current performance to the desired performance will require the same amount of time from beginning to end. So putting off the starting time simply delays the time when you will achieve your final objectives.

CHAPTER 3
THE PATH TO FAILURE

*I have not failed. I've just found
10,000 ways that won't work.*
Thomas Alva Edison

In the overall process of handling failures and their causes, a specific path to failure is assumed. It is necessary to characterize all failures using paths with common characteristics, so information can be organized in a consistent manner. On a number of occasions throughout this book, I have used an analogy with the path of a trip from one place to another to help describe how we use information concerning various points along that path, to provide insight concerning the overall path.

In this chapter, I will describe the overall path to failure and all the significant milestones along the way. Although you are likely to deal with hundreds or thousands of failure paths, all will share these same characteristics.

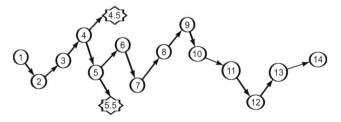

Assume that the diagram above charts the path of a failure from the beginning until the conclusion of the investigation. The numbers represent the following events:

1. Systemic Cause

A weakness exists in the organization or system. This weakness creates an opening that allows members of the organization to act in a manner that will result in the introduction of defects.

2. Human Cause

A specific individual either takes an action or fails to take an action that ultimately results in the introduction of a defect into an equipment item.

3. Physical Cause

A circumstance now exists that allows some form of deterioration to begin. This circumstance may be the leakage of water into an area containing two dissimilar metals, creating a "battery" and setting the stage for corrosion.

4. Failure Mechanism

A mechanism may exist that results in some form of deterioration. Most components are designed to be able to withstand some amounts of deterioration without being likely to fail. In describing a failure path, we are most often concerned with an unexpected failure mechanism (one that has not been considered in the design), or a failure mechanism that is working much faster than expected.

4.5 Once the failure mechanism is "working", there begins to be an opportunity to identify and use individual pieces of data and patterns that foretell an impending failure.

5. Defect

The deterioration resulting from the failure mechanism has proceeded to the point that a "failure-causing defect" now exists. Once the failure-causing defect exists there is a specific likelihood that a failure will occur. In many situations, the presence of the failure-causing defect simply introduces a statistical likelihood of a failure. When the luck of the draw places the defective component in a stressful condition, the failure will occur.

5.5 As with the failure mechanism, once the defect exists, it is possible to identify its presence through individual pieces of data or patterns that indicate that the risk of failure exists. Nature has begun throwing the dice and when the right combination of circumstances occurs, a failure will take place.

6. Failure

A failure has occurred. One of the critical functions for which the system or device is designed is no longer being done.

7. Malfunction Report

Someone issues a malfunction report describing the function that is impaired and the behavior it is presenting.

8. Diagnostics

After the malfunction report is issued, the individual assigned to take corrective action must begin trying to restore the function by finding the failed component and replacing it. The process begins by gathering external data that helps point to which system, sub-system, component, and condition can result in the specific behavior being experienced.

9. Funneling to Identify System, Sub-System

Funnelling is a part of the process of diagnostics. It consists of looking for clues that point to where the defect exists. In some instances the clues are subtle and in others they are highly prescriptive. If the person responsible has done a good job of mapping prior failures, a number of indicators might be recorded, so that they can be used to quickly identify a specific failure path and the right solution. If the person responsible has tracked the proportion of failures following each of a number of paths from the same starting point, it is possible to calculate the number following each

direction and therefore the relative likelihood of each.

10. Troubleshooting

Troubleshooting introduces the hands-on, invasive tasks needed to find defective components. If failure paths have been mapped in the past and the most likely path identified, the troubleshooting should be directed to begin with the most likely and proceed sequentially to the least likely. Since troubleshooting is invasive and often entails disassembly, it is possible that the troubleshooter will introduce defects during the process. As a result, it is best to be as effective as possible. An exception to the rule of starting with the most likely first is the instance in which one step might be far easier or less costly than others. It would then make sense to check that possibility early on the off-chance of quick success.

11. Identification of Failed Component(s)

When troubleshooting is complete, specific failed component(s) will be identified.

12. Identification of Component Condition Resulting in Failure

In addition to identifying the specific failed component, it is necessary that the aberrant condition be identified. If you cannot tell how the component has failed, you may not be able to tell if it has failed. If components are changed without determining the failed condition, the people changing them are just "parts changers". "Parts changers" may be providing short term relief, but they are not solving problems or helping improve performance.

13. Failure Analysis

Closely related but separate from identifying the failure condition is identifying the failure mechanism. For instance, a part might be broken in half. The break might have resulted from corrosion, erosion, fatigue, or overload. In each condition, the overall corrective action needed to prevent the same failure in the future will be different. This step is part of closing the loop with the beginning of the failure path by identifying one of the starting steps.

14. Root Cause Analysis

The final step in closing the loop is to identify the root cause that led to the failure. Most frequently, the three levels of cause (physical, human, and latent) become hidden in the process and bureaucracy of an organization. It is only by performing a thorough investigation and exposing the root cause that long-term improvements can be made.

When "X" Marks the Spot

One of the clichés associated with pirate adventure stories is the well-known fact that an "X" marks the spot on the map where the treasure is buried. Unlike the pirate business, the reliability business seldom has a treasure map leading to the one and only location where the solution can be found. There are numbers of possible "failure paths" and the diagnostician needs to determine which is most likely. Several failure paths begin with the same systemic cause. And as the analysis proceeds, each failure path can lead to a number of different failure modes.

There is a real value in understanding which paths are likely candidates and which are not. Although there is no clearly marked "X" to identify the correct path as on a treasure map, there are usually many clues that the diagnostician can use to help identify the true failure path.

- A clearly-written Malfunction Report (describing the specific function that has been impaired and the behaviors that are apparent), provides the best indication of the proper failure path.
- Occasionally there will be more than one "system" involved in a failure displaying specific symptoms. For instance, where there has been a total power loss, either the generator or the engine driving the generator might have failed. In this situation it would be best to provide some form of "discriminator" that would be helpful in pointing to the most likely culprit. For example, if the engine has been blowing black smoke or has thrown a piston through a crankcase wall, the indication would be an engine failure.
- Taking a step further, many systems are made up of a number of sub-systems. If it is possible to point to the affected sub-system, we can even more accurately define the failure path. An engine has several subsystems to be examined:

- Prime Mover
- Cooling system
- Fuel system
- Lubrication system
- Engine control

Some clues like loss of speed control or "hunting" may point to one sub-system as being more likely than others.

- Below the level of sub-system is the component level. A good practice is to review the history of the specific equipment item, and the history of similar equipment items, to identify which components experience most frequent or chronic failures.
- The last piece of information in the "visible" portion of the failure path is the condition of the failed component. If the objective is to eliminate future failures, the specific condition of the failed component is as important as any other piece of information. For example, four components may have failed. If one of them failed by corrosion, one by erosion, one by fatigue and one by overload, a single solution is not possible, and there are then four failure paths. If all four have failed by a single failure mechanism, there is a single failure path and a single solution may eliminate all the failures.

CHAPTER 4
LEARNING ABOUT A DEFECT

*Once the cat is out of the bag, you
need to think about how to deal
with a cat and not a bag.*
DTD

All too often, people think it is as easy to deal with failures
after they happen as it is to prevent them. These are typically not the people who have to don the bunker gear
and fight the fires. Failures follow their own chosen path
and, while valiant, our failure modes and effects analyses
are frequently only feeble attempts at describing the full
extent of reality. We can only guess at second- or third-
order effects. If a failure is allowed to occur, nature will
make the final choice where things will end.

You tend to learn about defects when the defect chooses
to make its presence felt. And defects always like to make
their presence felt at the least convenient times. So the
system we devise needs to account for the fact that failure causing defects are inconsiderate little buggers.

The first thing about a failure-causing defect is that it usually announces its presence to the least- articulate mem-
ber of any organization. As a result, the failure description
is likely to be difficult to understand and most likely incon-

sistent with any other failure that has ever happened in the past. Left to his (or her) own devices, the description of the resulting malfunction will be unlike all other reports and will be impossible to file or categorize.

 The most important point made in the previous chapter was that "there is nothing new under the sun". In other words, there are no new failures. Everything that happens has happened before (and is likely to happen again). As a result, it is possible to provide the individual who reports the malfunction with a choice of failure symptoms from which he (or she) can select the one that fits most closely.

You want to learn about the malfunction, but you want to learn about it in a useful manner. Correctly describing and classifying the malfunction in the first place will go a long way toward understanding the path the failure is likely to follow. In turn, that knowledge will go a long way toward solving the problem and eliminating the defect.

So just what is a good malfunction report?

Every critical system performs one (or more) important "functions". A function is the reason for its existence. If the function did not need to be accomplished, the system or device would not exist and you would not spend money to operate or maintain it.

Function:
The action or service performed by a system or device. The conversion accomplished by a system or device is best described by comparing the characteristics of the out-going product to that of the in-coming.

So, the first part of a malfunction report is a description of the function that is behaving badly.

The second part of a malfunction report is a brief but complete description of the behavior that does not meet requirements. It is important to understand the "behavior that completely meets requirements" because a departure in any manner can be viewed as a "malfunction".

 Let's use the example of a steam whistle. A steam whistle is a warning device whose function is to provide a warning for one reason or another. In a plant or on a system, there might be a number of warning devices and there might be a number of steam whistles performing various functions. Maybe this specific steam whistle is intended to sound the alarm for fires, so it becomes the fire whistle. The requirement is that there is no confusion. Whoever creates the malfunction report and whoever reads it should have absolutely no confusion over the specific problem that is being indicated and the specific device that is causing it. In some plants, every equipment item is numbered, which helps in our project.

On one occasion when working for a large company with many plants, I was asked to visit a plant that had been newly acquired. In the system with which I was familiar, all pumps had equipment identification numbers that had the general format P-12345. At this newly acquired facility, pumps had no equipment numbers. Instead, they had names or colors. The work orders would ask for maintenance on "Big Blue" or "Old Shakey". This approach worked in a paper work order system and with individuals who were familiar with the plant, but it was a problem for a computerized maintenance management system (CMMS) and for individuals who were unfamiliar with the plant.

It is important to identify all the behaviors that describe the possible symptoms. Individuals filling out the malfunction report can then be asked to select from one of the realistically possible alternatives.

For example, a steam whistle has been known to misbehave in the following ways:

1. Does not sound
2. Sounds weakly
3. Sounds intermittently
4. Sounds continuously (Does not shut down)

It is important to avoid asking the person reporting the malfunction to go beyond what they actually know. For instance, if the sound coming from the whistle is weak, the person should not be asked to comment on the condition of the valve that controls the steam. That is beyond the knowledge of the person making the report and will just add confusion. On the other hand, if the person reporting the malfunction has access to a pressure gauge telling the steam pressure going to the valve, it would be fair to ask for that information. Knowing that the steam pressure is less than normal might explain the cause without going any further.

> **Malfunction:**
>   The situation when the service or transformation performed by a system or device no longer meets expectations, or when the output no longer meets requirements. A malfunction is best described by indicating which function is affected and the specific behavior that is not meeting requirements.

It should be emphasized that the best reporting systems are those that are "interactive". By that I mean the individual(s) using the system will both be asked questions by the system and will be able to access information in the system to answer his or her questions. (In other words, they will know what failure paths are realistic and possible.) With this additional requirement, we will create a system that asks questions to help either focus on most likely causes, or eliminate unlikely causes. In addition, the individual using the system can "peek" at reasonably possible scenarios for clues to the right answer.

For instance, assume that a specific symptom leads to only two possible failure modes: A and B. If failure mode A has a specific characteristic that B does not, it is impor-

> **Failure Path:**
>   A failure path is the sequence of events and conditions, beginning with precursors or clues that a defect is forming, continuing through deterioration and the development of the defect, and leading to a failure. In the context of comprehensive corrective action systems, this path will continue with identification of the failure mode, the failure mechanism, and the several levels of root cause.

tant to determine if that characteristic is present. The malfunction notification system you create should do two things: first, it should ask the questions necessary to distinguish between failure mode A and failure mode B. Second, it should make descriptions of both failure modes available to the user so that he or she knows what is possible.

 A variety of additional characteristics frequently accompany a failure that will help determine how the triage, troubleshooting, and repair should proceed. Modern computer-controlled systems might provide additional data that will help point the diagnostician in the right direction. Modern computer-controlled systems frequently have the ability to generate fault codes when something is operating outside an acceptable range. Occasionally, the problem is even more apparent. If a device is operating below freezing temperature, that fact might be the cause. If a reciprocating engine breaks a connecting rod and throws a piston out, the problem is readily apparent.

When specific supplemental information can be associated with specific malfunction descriptions, it should be automatically requested by the system. For instance, in an automated database, when the individual identifies a malfunction report (or malfunction code) that requires additional information to provide insight and direction, the appropriate question should automatically "pop-up".

Pop-Up Box:
   A pop-up box in a computer database is a field that has been populated with a pre-determined group of choices that appear when the field is selected. The objective is to prevent the field from being used as a free-form field and to cause all choices to be "identified" in a specific manner.

 Let's assume that someone has completed a malfunction report and that malfunction report matches one of the known functions and behaviors. Let's further assume that either the system has automatically asked, or a human participant has provided, all the supplementary information needed to determine which failure modes are possible and which are not possible. In other words, we have a fairly clear understanding of the starting point for this path and that starting point is something with which we are familiar.

Once a malfunction report has been issued, someone

must begin thinking about how to deal with the problem. As the saying at the beginning of the chapter stated, once the cat is out of the bag, you need to start thinking about how to deal with a cat and not a bag. It would have been easier to prevent the problem, but now there is a malfunction and the malfunction suggests there is a defect in your equipment. Thinking is less costly than jumping in and starting to disassemble things, so the best first step is to learn as much as you can about the possible defect. The person assigned to learn about the defect is a diagnostician.

A diagnostician typically has a variety of tools available.

- The first tool is the diagnosticians own experience. Many diagnosticians have worked as mechanics or foremen, and have had some experience diagnosing problems. He (or she) should call on that experience.

- If you have been doing a good job of mapping the specific symptoms associated with various forms of malfunction reports to the possible failure modes that can realistically cause those symptoms, the diagnostician can use the database.

- Many companies do a good job of keeping records of maintenance failure history of systems and equipment. The history should be referenced to see if there is a chronic problem that appears to have been left unsolved in the past.

Once the diagnostician has referenced the information

and feels adequately prepared, he (or she) will provide instructions for some form of action. At this point, the action should still be viewed as a step in learning more about the defect rather than providing a permanent solution. The following actions are possible:

1. Instantaneous Repair – The last chapter described an example of a quick-fix involving recycling a computer. While this example and other quick fixes might provide temporary relief, they typically do not eliminate the inherent defect. If the quick fix works, it is important to record that fact. The fix might be used again in the future and it might provide some insight into the cause of the defect and the ultimate solution.

Triage:
Triage is the act of determining the order of in which assistance should be given, based on the likelihood of success and the severity of the condition.

2. Triage – Many organizations have several ways they can repair things. The approaches vary in the amount of time they take to complete repairs and they vary in the level of complexity. As variety and complexity increase, so does the time required to repair. In conducting triage, the diagnostician will want to send simple problems to fast-track problem solvers and complex problems to problem solvers better suited to more difficult problems.

3. Order of troubleshooting – If your system has been properly mapping malfunction systems to final failure modes for some time, the statistics will tell which failure mode is most likely, which is second most likely, and so on. The activity of troubleshooting often includes some amount of disassembly and testing, so the cost of labor is involved. The objective should be to minimize the cost of labor so it is important to pursue most likely alternatives first then move sequentially to the least likely.

When troubleshooting is complete, the troubleshooter will know the Failure Mode. At this point, the troubleshooter will either repair the problem or write a detailed work order describing the tasks needed to perform the repair. The troubleshooter will also document the failure mode in the record keeping system. By recording the specific component that failed, and the condition of the component that led to the failure, the troubleshooter is adding information to the database that will make future diagnostic and repairs more efficient and effective.

Although the immediate problem may be solved, we still want to learn more about the defect for several reasons.

First, we want to learn if the quick-fix worked. If it did, that is good; it can be used again in the future. On the other hand, as mentioned earlier, the quick-fix probably didn't eliminate the defect. If we want to eliminate the defect and rid ourselves of the nuisance, we need to record and report the characteristics associated with the failure, including behavior and any data showing on computer screens.

Second, we want to identify the cause of the problem. There are two further steps in finding causation. One is failure analysis and the other is Root Cause Analysis.

Failure analysis will identify the "failure mechanism". Unlike the "failure mode" there are only a few failure mechanisms. As an example, let's assume that the failure mode is that a compressed air supply line is plugged. The failure mode would read "air supply line – plugged". In performing the troubleshooting, the technician may find magnetic corrosion products in the plugged line. In this event, the failure mechanism would be "corrosion".

Learning about the failure mechanism is important to understanding and eliminating the cause to make sure that the failure does not recur. It may be possible to solve the current problem without looking further, but that approach is short sighted. It would be far better to understand the failure mechanism that produced the deterioration that caused the failure. Even further, it would be better to follow up on the root cause that allowed the deterioration mechanism to proceed to failure.

Root Cause Analysis (RCA):
  RCA is a form of detailed, structured, and detailed analysis that identifies the cause or causes that allowed a defect to form and a failure to occur. The elimination of this cause or these causes, will prevent formation of future defects of this kind and related failures.

In our example, the failure mechanism is corrosion. If the air line concerned is a part of a compressed air system, the air in the system needs to be dry to minimize the possibility of corrosion.

In the example, having found corrosion in the system, we should pursue the problem further. If we find liquid water in the system, we might go back and check the compressed air dryer.

If the air dryer was not functioning and was allowing moisture into the system, this moisture would be the "physical cause". The physical cause is the first level of root cause analysis.

The next question is, "why is the air dryer not working?" Air dryers do not operate, maintain, or inspect themselves, so there was a human who did not fulfill his (or her) personal accountability to see that the air dryer was in proper operating condition. Maybe the name of the accountable individual is "Joe". Joe is the human cause of the failure.

But everybody knows Joe is a good person. Joe doesn't make mistakes on purpose. In fact, very few people do. In most such events, there is a latent or systemic cause for the problem. Maybe Joe wasn't properly trained. Maybe he didn't have the time. Maybe he didn't know he was responsible for ensuring that the air dryer was maintained.

When you know why Joe made the mistake you will know the latent or systemic cause. Once the systemic cause is known, it is possible to turn this failure mode off like a light switch. The same systemic weakness is frequently responsible for more than one kind of failure, so correcting a systemic weakness may eliminate a whole variety of failures.

CHAPTER 5
MALFUNCTION REPORTING

*The right to be heard does not automatically
include the right to be taken seriously.*
Hubert Humphrey

I recall a situation more than twenty years ago when it was commonplace to simply write out "work orders" to specify what needed to be done. One such work order asked for the replacement of some 180-foot of welded steel condensate piping spread across fifteen stories of a structure. The first attempt required about a week. Shortly after the first attempt, it was determined that the desired

---

Work Order:
A work order is the element of a work management system that is used to:
- Identify the problem and the task(s) needed to correct it.
- Obtain approval from appropriate authority for expending the resources being requested.
- Identify the specific equipment involved in the failure.
- Identify the location.
- Identify the proper accounting codes.

When viewing the part a work order plays in the reliability process, it may also be the interface with database(s) that contain:
- Malfunction report
- System/Equipment items involved
- Failure Mode

objective of the work order was not achieved so a second attempt was initiated. After that a third and a fourth and a fifth attempt was made. In all, seven attempts were made. People were busy and work was being done three shifts a day and seven days a week so it was easy to lose track of a job that was thought to be very simple. Finally, someone noticed the continual draw on resources and asked what the person who initiated the work order was trying to achieve. It turned out that the line was plugged in a portion of the piping system close to the bottom. This portion of the system had never been changed in any of the seven attempts.

Had the request identified the problem that needed to be solved rather than describing a solution, a great deal of time and money could have been saved. The individual who wrote the work request thought he knew what was wrong. He also envisioned a response to the work request that would address the problem. Neither of these perceptions was correct. The person who issued the work request did not know where the pipe was plugged nor how the maintenance crew would respond.

A better approach would have been to create a "malfunction report". As the name implies, the malfunction report is used to identify a function that is behaving in an unacceptable manner and the specific behavior. That description probably seems more complex than it is.

To begin with, each system performs one or more critical functions. You perform maintenance to ensure that those functions are preserved. If a service other than a critical function is halted, you are not as concerned with restoring operation as you are for critical functions. So a malfunction report begins by identifying the "function" that has been interrupted.

Next, each function has specific modes of "behavior" that meets requirements. When the system is behaving in a manner that does not meet designated requirements, there is a malfunction. So the second part of a malfunction report is the clear description of the unacceptable behavior.

It is important to clearly identify the affected function and the behavior, and it is also important to avoid identifying too many shades of each behavior. In other words, something is on, off, or somewhere in the middle. The "somewhere in the middle" either meets requirements or it does not. There is only value in differentiating the "shades" of performance when the distinctions will lead to different possible failure paths and different solutions.

Let's use another analogy to describe this point. Earlier we used the analogy of flights from an airport to direct connections. Let's continue with that analogy.

Say we are traveling from Chicago. There are two large airports in Chicago, O'Hare and Midway. Chicago O'Hare Airport serves a large number of direct destinations. Midway serves a smaller number of direct destinations and has some overlap with O'Hare. If I knew a passenger was leaving from O'Hare, I would also have knowledge of his (or her) possible destinations. Based on ridership to each direct destination, I would also know the relative likelihood that he (or she) is headed to each one. The same is true for a passenger leaving Midway. But for Midway, the destinations being served and the rider- ship is different so the results of the analysis would be different.

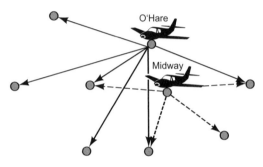

In this instance, there is more information in knowing that a passenger is leaving from either O'Hare or Midway than simply Chicago. As a result, it would be better to add that detail.

The same is true for mapping failures from two malfunction reports (function-behavior) that are very similar but different in some key way. Understanding the differences will allow the diagnostician to recommend different solutions. Recording the differences will allow us to track the differences in "destinations" and then calculate the likelihood of each.

On the other hand, if the starting point is effectively the same, and all destinations are the same, there is no value in creating unrealistic distinctions. That action will only result in separating failure paths into two (or more) separate areas, and reduce the focus on a single source of problems.

Let's discuss a few examples of a malfunction report. Many facilities have one or more kinds of warning devices. In each of these situations, the function is "warning". But since there may be more than one warning device it is important to be explicit: Warning – Horn, Warning – Bell, Warning - Siren. Each of these mechanisms is made up of different components and will have different failure paths.

If we select the horn, it can have a variety of behaviors that can be called malfunctions. For instance, the horn could not sound, it could sound intermittently, it could sound with a weak tone, or it could stick on and sound all the time. Combining each of these with the associated function creates a distinct malfunction report.

- Warning Horn – Fails to Sound
- Warning Horn – Sounds intermittently
- Warning Horn - Sounds in a weak manner
- Warning Horn – Will not shut off

Thinking about describing these malfunction reports in a manner that is easy for a database to recognize, use, and sort, we can use alpha or numeric characters to represent each portion of the description. It is important to keep the ultimate size of your system and the number of functions and behaviors in mind. If there are more than 99 func-tions, you will want to use a three-number or an alpha code for the function. The same is true for the behaviors.

For simplicity, we will select WH to represent the warning horn function and the numbers 01 thru 04 to represent the four behaviors.

- Warning Horn – Fails to Sound   - WH01
- Warning Horn – Sounds intermittently   - WH02
- Warning Horn - Sounds in a weak manner - WH03
- Warning Horn – Will not shut off   - WH04

In performing the diagnosis, additional information might be important. For instance, if the horn is air powered and the ambient temperature is below freezing, moisture in the air may freeze and cause the malfunction. If the horn is steam powered and the steam pressure is below nor-mal, the low steam pressure might be responsible for the malfunction.

The importance of creating additional coding to account for supplementary information would depend on how important those features are to deter-mining the cause of failure and the fail-ure path. In some situations, weather extremes (hot, cold, wet) have a signif-icant impact on failures and therefore should be recorded. If measures have been implemented to mitigate the impact of weather extremes, failures dur-ing periods of severe weather would suggest that the measures are not working.

Other supplementary information can include "fault codes" such as those provided within computer control systems. In addition to specific translatable fault codes, screens can go completely black, or can show all stars, or all dashes. These displays have specific meanings, and capturing the information from the display with the Malfunction Report can more accurately describe the starting point of the failure path.

Fault Code:
Computer-based control and condition monitoring systems frequently use fault codes to portray warnings that a specific event has occurred or a certain condition exists. Fault codes are alpha-numeric codes that represent a much longer description of conditions or events. Fault codes frequently need to be sorted out from "Incident Codes" or simple "state changes," that may be reported by these systems but represent no unusual condition or impending failure.

CHAPTER 6
DIAGNOSTICS

It is a characteristic of wisdom not to do desperate things.

Henry David Thoreau

For reliability engineers, television programs describing emergency room procedures contain more than simply entertainment value. These programs use emergency response jargon that is directly applicable to identifying the defects in an urgent yet organized manner.

Let's begin with the term triage. In an emergency room situation, the typical "first come, first served" approach is ineffective. Although a broken arm might be painful, it can afford to wait while a heart attack victim is being treated. In a similar manner, some functions and some behaviors are more important than others. All malfunctions announce themselves in one manner or another, but cool and calm situation assessment is needed to determine which needs immediate response and which can wait. That cool and calm assessment is described by another medical term, diagnosis.

In medicine, diagnosis refers to the act of identifying the disease or affliction. If we analyze the steps a medical doctor takes in developing a diagnosis, we see that the pattern is relatively fixed. In fact, routine medicine is more

of an art than a science. Medical professionals study science so that they understand the background of the steps they take. But when the doctor takes those steps he (or she) is simply following a pattern. A "good" doctor is one who selects the best pattern to follow most quickly.

Let's take a few steps back and see what makes a doctor a "good" doctor. First, a "good" patient makes a "good" doctor. If the patient is sensitive to his (or her) own symptoms and reports them in a thorough and accurate manner, the doctor is far more able to make a good diagnosis. Second, a lot of applicable experience makes a "good" doctor. If the doctor has seen your symptoms repeatedly in the past, and has similar experiences in following the symptoms through to cure, it is likely that the doctor will seem quite capable to you. Third, if the doctor has had a wide variety of experiences, he will be able to deal successfully with a wide variety of illnesses.

One thing that is taken for granted is a good memory. Much of a medical doctor's education is simply memorization of a lot of information. In addition, a medical graduate has proven that he (or she) can not only remember things, but that those things can be recalled and applied when needed. Throughout his (or her) career, every doctor will need to apply that facility for memorization and recall every day.

When a doctor provides a diagnosis for an illness, he (or she) can fairly well describe the entire path the illness will follow. Clearly there are subtle differences from patient to patient in strength, immunity level, attitude, and a variety of other factors that affect the path to recovery. But generally, once the disease is recognized, the doctor can describe how the disease and recovery will progress.

The ability to do this diagnosis routine is not based on clairvoyance. It is based on experience and the ability to catalog and recall similar experiences.

Diagnosing the physical defects that led to equipment failures can follow much the same pattern as diagnostics performed by medical professionals. Diagnosis is not troubleshooting. The process of diagnosis may include a little poking and prodding but generally does not include any invasive procedures. For the most part, diagnoses are achieved by reviewing external measures, conditions and behaviors. As with the medical diagnosis, if records are comprehensive and detailed and if the system storing the records makes them readily accessible, the diagnoses are likely to be quick and accurate.

The role of diagnostician of a failure within a physical system is like being the individual who calls the plays for a football team. That person can be the coach or the offensive coordinator but is seldom the quarterback. The person calling the plays knows the downs and distance to a first down. He knows the position on the field. He knows the plays in the playbook that are designed for this situation. He knows the plays that have been successful so far in the game and the tendencies of the opposing defense. From all that information, the person calling the plays is able to select a play that is most likely to succeed.

Similarly, the diagnostician is the person who determines what to do next. Sometimes it will be troubleshooting (just trying to maintain possession of the ball), and sometimes it will be the ultimate repair (going for a touchdown).

 Many maintenance organizations do not specifically include a person with the title diagnostician, but it is a function that always happens, even if by default. Someone always has to make a choice on how to approach the repair. The individual performing that function may only be qualified to guess. If so, the efficiency of the repair method will be based on guesswork. If the person performing the diagnostics has some experience, the diagnostics will be based on that experience. If the person has access to a comprehensive well-disciplined database, it is more likely that the diagnosis will lead to an accurate solution in the fastest way possible.

When assembling a database, the information included in the diagnostic portion can be the most difficult to gather. The diagnosis connects the malfunction report to the failure mode. The malfunction report is a description of the function experiencing the problem and the way it is currently (mis)behaving. The failure mode is a description of the specific component that is defective and the condition that constitutes a defect. The diagnosis is a critical activity that "maps" the path from the reported malfunction to the component or components that can be causing that malfunction being experienced.

For instance, think of a simple air horn system. Its function is to provide an alarm. The air horn system is composed of three subsystems:

1. The subsystem that conveys the air from the compressor
2. The subsystem that switches the horn on and off
3. The horn itself

The subsystem that conveys the air is made up of three components:
- The tubing from the air supply header to the solenoid valve
- The solenoid valve
- The tubing from the solenoid valve to the horn

The subsystem that switches the horn on and off is made up of three components:
- The conductor (wire) from the power supply to the push button
- The pushbutton
- The conductor from the pushbutton to the horn

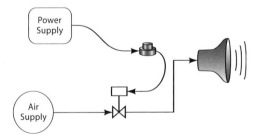

The function of this system is to provide an alarm, but there are several ways it can behave that would require a malfunction report to be filed. The horn:

1. can simply not sound.
2. can sound intermittently
3. can sound in a weak manner
4. can be stuck on and sound all the time

Emergency Communications - Air Horn System			
Function	System	Subsystem	Component
Emergency Communications - Horn			
	Air Horn		
		Air Supply	
			Tubing from Air Header to Mag Valve
			Mag Valve
			Tubing from Mag Valve to Horn
		Switching Circuit	
			Wire from Power Supply to Pushbutton
			Pushbutton
			Wire from Pushbutton to Mag Valve
		Air Horn	
			Air Horn

The specific behavior will provide insight into which of the subsystems can be involved in the failure. For instance, if the horn does not sound, any of the subsystems can be at fault. On the other hand, if the horn is sounding all the time or even intermittently, the horn itself has to be excluded from consideration. If the horn sounds weakly, but always responds to the pushbutton, the fault is likely not to be the switching circuit.

When the horn does not sound at all, the database of historical records will point to the need to diagnose all three systems. If historically 70% of the failures are in the switching subsystem, and the majority of those are loose connections at the pushbutton, a relatively simple repair would be indicated. The diagnostician would recommend that the push button be inspected and the wire connectors tightened. If 25% of the failures were in the air supply system and the final 5% of failures were in the horn, these subsystems should be troubleshot second and third.

The comprehensive map from each and every Malfunction Report code to each appropriate Failure Mode code provides the diagnostician with both the possible candidates and the likelihood of each candidate. This information is like the playbook for the individual calling plays in football.

Depending on how your organizational roles are assembled, it may be possible for the person receiving the malfunction report and performing initial diagnosis to ask for additional information or provide simple instructions. For instance, if the horn is stuck on, is it possible for some-

Emergency Communications - Air Horn System

Function	System	Behavior	Subsystem	Component	Failure Mode
Emergency Communications - Horn	Air Horn				
			Air Supply		
		1,3		Tubing from Air Header to Mag Valve	Tubing Plugged
		1,2,4		Mag Valve	Valve Sticking
		1,3		Tubing from Mag Valve to Horn	Tubing Crushed
			Switching Circuit		
		1		Wire from Power Supply to Pushbutton	Wire Worn Thru
		1,2		Pushbutton	PB Worn Out
		1,2		Wire from Pushbutton to Mag Valve	Wire Loose Connection
			Air Horn		
		1		Air Horn	Air Horn Missing

Behaviors

1	Air Horn cannot sound.
2	Air Horn sounds intermittently.
3	Air Horn sounds in a weak manner.
4	Air Horn is stuck on and sounds all the time.

one to tap the pushbutton or the solenoid valve. If either is stuck open, tapping might release them. If tapping cannot provide a temporary solution, having unsuccessfully completed the act provides more useful information.

If your organization is set up to provide several levels of repair geared to various complexities, knowing which subsystems are involved will help determine where repairs should be accomplished. Some organizations based on "Total Productive Maintenance" have provided skills and training for the equipment operators so they can make basic repairs on their own. If a DIN (do-it-now) crew carries a supply of switches, solenoid valves, and air fittings, almost any repair short of a horn replacement could be quickly accomplished. Some organizations separate repairs made by skilled crafts in the field from those made in shops. All in all, the closer the repair can be made to the point of failure, the better. Quick repairs with the least costly resource performing the repair will result in the lowest cost and the smallest amount of downtime.

If the diagnostician has an intimate knowledge of how, where, and what repairs can be accomplished, and knows the likely repairs that will be required, he (or she) can both perform triage and provide instructions to the troubleshooter. This knowledge is like knowing down and distance, field position, and available plays on a football field. By combining this knowledge with the map of Malfunction Reports and probable Failure Modes, the diagnostician can make the most logical call.

DIN – Do-It-Now:
  Occasionally, organizations will separate the maintenance workforce into two parts. One part works exclusively on planned and scheduled work, and therefore can improve efficiency and effectiveness because of that. The other part works on emergencies when there is emergency work to do (and non-emergency work when there is a lull). The DIN or emergency crew carries the kinds of tools and materials generally needed to facilitate repairs that can be done quickly, but still properly.

CHAPTER 7
TROUBLESHOOTING

Shallow men believe in luck.
Strong men believe in cause and effect.
Ralph Waldo Emerson

For purposes of our discussion, let's distinguish diagnostics from troubleshooting. We will assume that diagnosis involves analysis of externally-available information, and troubleshooting may involve physical contact and disassembly. It is possible that someone gathering diagnostic information might come in contact with an equipment item to observe externally visible conditions, but that contact is not the same as troubleshooting.

Further to the horn example from the last chapter, horns frequently grow legs and walk away (are stolen). As a result, when the horn doesn't work, it is a good idea to start by determining if the horn is still there. If not, you will know the failure mode - "Horn - Missing". If the horn is still there, it would be useful if the operator can check to see if the air system is functioning at normal pressure and that the electrical system has power. Again, these checks do not constitute troubleshooting; they provide additional information to help form the diagnosis. If the tubing that provides air, either to the solenoid valve or to the horn is in plain sight, it would be useful for operator to look them over to see if they are crushed or broken. If the wire to the solenoid is exposed, it would be helpful if the operator can see if it is cut or grounded. If this kind of information is unavailable, it will be necessary for the troubleshooter to start by gathering it. If the information is available, it will help form the diagnosis and allow the troubleshooting to be accomplished much more quickly.

Assuming that you have established a system that:

1. Clearly identifies each discrete Malfunction Report (Function and Behavior),
2. Clearly identifies each Failure Mode (Component and Condition),
3. Maps each applicable Failure Mode to the appropriate Malfunction Report and
4. Has accurately bucketed all failures and repairs for several years;

The diagnostician will be able to direct the repair to the personnel most likely to be capable of performing the repair and to identify the sequence in which troubleshooting should be accomplished.

For example, say that the following is the order in which repairs have been required in the past:

1. Switching Circuit - 70%
   a. Electrical supply from pushbutton to solenoid – worn through – 35%
   b. Pushbutton – worn out – 25%
   c. Electrical supply from power supply to push button – loose connection – 10%

2. Air Supply – 25%
      a. Solenoid Valve – Sticking – 10%
      b. Air supply to solenoid valve – plugged – 10%
      c. Air supply to horn – crushed – 5%

3. Horn – 5%
      a. Horn – missing – 5%

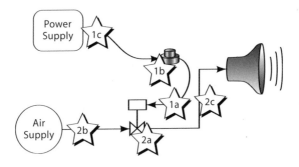

The diagnostician would likely triage the repair to personnel who are capable of minor repairs including electrical work. In providing a troubleshooting recommendation, the Switching Circuit should be analyzed first, specifically looking for grounds in the electrical supply lines. If the individual performing troubleshooting can also make minor repairs, it would be best if he (or she) carry some replacement wire and a replacement pushbutton. These small materials would address 70% of possible failure modes.

In providing advice to the troubleshooter, the diagnostician would recommend first looking for grounds in the wire between the pushbutton and the solenoid. Second,

Emergency Communications - Air Horn System

Function	System	Behavior	Subsystem	Component	Failure Mode	Likelihood
Emergency Communications - Horn	Air Horn					
			Air Supply			25%
		1, 3		Tubing from Air Header to Mag Valve	Tubing Plugged	10%
		1,2,4		Mag Valve	Valve Sticking	10%
		1, 3		Tubing from Mag Valve to Horn	Tubing Crushed	5%
			Switching Circuit			70%
		1		Wire from Power Supply to Pushbutton	Wire Worn Thru	35%
		1,2		Pushbutton	PB Worn Out	25%
		1,2		Wire from Pushbutton to Mag Valve	Wire Loose Connection	10%
			Air Horn			5%
		1		Air Horn	Air Horn Missing	5%
Behaviors	1	Air Horn cannot sound.				
	2	Air Horn sounds intermittently.				
	3	Air Horn sounds in a weak manner.				
	4	Air Horn is stuck on and sounds all the time.				

the troubleshooter should check for a worn pushbutton and third, make sure that all connections are tight. (In real life, the troubleshooter would likely give all the wires a tug first because that is the simplest way to check.)

This approach may seem somewhat trivial when dealing with such a simple system, but a lot has to do with scale and logistics. If the components are hundreds or thousands of feet apart, a little knowledge could save a lot of time. Also, if the storeroom is some distance from the work site, carrying parts and materials likely to be needed could prevent multiple trips.

There are other ways to calculate the proper inventory levels, but knowing specifically where parts and materials are being used may make it possible to create satellite warehouses to reduce transit time.

When the troubleshooter concludes his (or her) work, there will be two products. The first product will be an identified failure mode. The failure mode will identify the specific component that is defective and the condition of that component that led to the malfunction.

The second product will be one of two possible alternatives. The first alternative is a repair and the second product is a work order describing the specific task needed to complete the repair. In some situations, the troubleshooter can complete repairs and in others, crafts persons must be assigned to complete repairs. (In some shops, the crafts persons are also troubleshooters.)

For purposes of analyzing and improving reliability, our

primary interests rest with the first product, the Failure Mode. It is important for the troubleshooter to describe the specific component being used and the specific condition of that component.

CHAPTER 8
DIGRESSION CONCERNING FACTS

*People always seemed to know half of history,
and to get it confused with the other half.*
Jane Haddam

The successful pursuit of reliability demands that we all deal in "facts". Facts play a part in two hierarchies that are critical to analysis.

The first hierarchy begins with "data" and works its way upward in terms of utility. This hierarchy goes as follows:
- Data
- Information
- Knowledge
- Wisdom

Starting with the definition of "data" and continuing on:

Data – Information in the form of **facts** or figures obtained from direct observations, used as a basis for making decisions or drawing conclusions.

Information – The comprehensive collection of **facts** and data about a particular subject.

Knowledge – Familiarity with, and understanding of, information concerning a particular subject gained through study and experience.

Wisdom – The knowledge and experience needed to make sensible decisions and judgments. The good sense shown by decisions and judgments made in the past.

When we are asked to identify a root cause, to create a diagnosis, or to solve a problem, we are being asked to exercise our "wisdom," and wisdom has its basis in data. And finally, to be useful, data must have its basis in fact.

In a high-volume, high-speed environment; we must depend on the computer to perform a significant portion of the analysis and data reduction. As a result, the data we use must not only be factual, it must be recognizable by the computer. Simply described, we need to create "buckets" that are intended to contain all of a generic category of malfunction and all of a generic category of failure mode. The incidents falling into a single "bucket" do not have to be exactly the same; they simply have to be similar enough to fit in the same "bucket". All incidents falling into the same bucket need to be described using the same code, which can be recognized and sorted by the computer. A long narrative description does not provide usable data.

Examples of two failure modes that fall into the same "bucket" are:

Horn air line plugged – (In this example, the air line from the compressed air header to the horn is plugged with dirt at a location close to the horn.)
Horn air line plugged – (In this example, the air line from

the compressed air header to the horn is plugged with corrosion product at a location in the middle of the run to the horn.)

These two situations will ultimately map to different failure mechanisms, and they will be corrected using similar procedures.

Examples of four failure modes that do not fall into the same "bucket" are:

1. Horn air line plugged – (Here, the air line from the compressed air header to the horn is plugged with dirt at a location close to the horn.)
2. Horn air line cracked – (Here, the air line from the compressed air header to the horn is cracked and losing air at a location close to the horn.)
3. Horn air line crushed – (Here, the air line from the compressed air header to the horn is crushed at a location close to the horn.)
4. Bell air line plugged – (Here, the air line from the compressed air header to the bell is plugged with dirt at a location close to the bell.)

To support accurate diagnosis, improve efficiency of repairs, and identify components with chronic problems, it is important to assign the code associated with the correct failure mode. In the reliability business, these capabilities constitute "wisdom".

The second hierarchy involving "facts" begins with the term "fact" and works its way downward in terms of accuracy and utility.

Facts – Pertinent, precise, accurate, verifiable, specific, measurable, "hard **data**"

Inferences – Logical deduction based on facts

Assumptions – Logical hypothesis (which if true) could explain the facts

Opinions - Based on memory, experience and intuition

Beliefs – Based not on personal observations, but on the opinions espoused by others

Hearsay – Second-, third-, or fourth-hand information, generally not attributable or verifiable

Guess – Possibly educated, but not based on data or analysis

Fantasy – No basis, possibly a distortion

When we are asked to verify that some data is a fact, it is useful to begin by comparing the data to the definitions provided above.

Have you personally seen it or just heard it from someone else?

If it is hearsay, is it attributable to a specific person so that you can verify the accuracy? Or is it lower on the list … like a guess or fantasy?

If it is an opinion based on memory or experience, was that memory or experience based on fact? Did you verify the fact at the time?

If it is an inference or assumption that tends to fit the facts, how many other inferences or assumptions could also fit the facts?

The best rule to use when performing an analysis or inputting data is, "**Just the facts**." That means precise, accurate, verifiable data that has been gathered first hand.

One of the best tools to use in collecting facts is the "5-Whys". A rule-of-thumb is that by the time you have asked "why" for the fifth time, you will have reached a factual account.

An example of a 5-Why script follows:

• Why do you think the pump failed"
        Because the bearing burned up.

• Why did the bearing burn up?
        Because the pump overheated.

• Why did the pump overheat?
        Because the pump was being operated with the suction valve closed.

• Why was the suction valve closed?
        Because the operator did not line it up properly.

- Why didn't the operator line up the pump correctly?
  Because he was in a hurry to make a shift
  change and leave for the day.

CHAPTER 9
FAILURE ANALYSIS

Things should be made as simple as possible, but not any simpler.
Albert Einstein

When the troubleshooter concludes his or her work, the Failure Mode will be known, but not the Failure Mechanism or the cause. There are a myriad of Failure Modes, but there are just a few Failure Mechanisms.

If it is a mechanical component, the only possible Failure Mechanisms are:

- Corrosion
- Erosion
- Fatigue
- Overload

Several kinds of corrosion are possible and frequently the presence or absence of corrosion can be determined by several elements:

- The presence of a cathode (more noble metal)
- The presence of an anode (less noble metal)
- The presence of an electrolyte (liquid completing a circuit between the cathode and the anode)
- The presence of corrosion product (rust or other metal oxide)

Erosion or abrasion will typically leave wear marks and products of wear.

Fatigue is the process of going from tension to compression in a repeated manner over an extended period of time. The severity of bending needs to be sufficient to result in stress levels above that required to cause damage. The number of cycles needed to cause failure is dependent upon the stress level. Given enough time, even fairly small stress levels can result in failure.

In many situations, vibration resulting from inadequate support will result in fatigue failures over time. In other examples, misalignment or imbalance of rotating components can cause fatigue and ultimate failure.

Overload is the result of a component being designed for a smaller load than it is being asked to carry. The overload can be the result of under-design or proper design and over-loading. It is always necessary to compare the actual load to the design load to determine if overload is present.

In the case of an electrical component, the Failure Mechanisms are different but analogous to the mechanical failure modes:

- Overload – Supply Transient
- Overload – Load Stall
- Electrical Equivalent of Fatigue
- Insulation Breakdown due to Heat
- Insulation Breakdown due to Chemical Attack
- Mechanical Abrasion
- Mechanical Loosening

Many of these mechanisms are self-explanatory, but several are worth discussion.

Electrical overload can result either from a transient variation in the supply causing an over-current condition, or from the load stalling and the normal current turning into heat.

The electrical equivalent of fatigue is caused by repeated exposure to conditions that are beyond the capability of the system but are insufficient to cause instantaneous failure. An electrical device can be exposed to supply transients, load stalls, extremes in temperature, or a combination of several of these events that will not result in instantaneous failure but will cause deterioration that will lead to failure over time.

Insulation can break down for a variety of reasons including: exposure to temperatures beyond the material capabilities, exposure to chemicals that are incompatible with insulation, and exposure to ultraviolet radiation.

Probably the most common Failure Mode for electrical equipment is "mechanical failures," or failures that affect mechanical aspects of electrical systems. In situations where designers have not been careful to minimize connections and to select the proper kind of connections for the service, mechanical loosening leading to poor contact is a common failure mode.

Identifying the Failure Mechanism is an important step in the path to identifying the root cause. Failure Mechanism is another element that can be easily characterized by a code. Sorting by Failure Mechanism code is then helpful in identifying common or chronic problems. For instance, if corrosion is a common failure mechanism for mechanical components in a specific facility, it may be helpful to see if the painting and lubrication programs are adequate. Identifying "chronic" failure mechanisms can be helpful in preventing seemingly unrelated failures.

CHAPTER 10
ROOT CAUSE ANALYSIS

*If you have always done it that way,
it is probably wrong.*
                    Charles Kettering

The terms "Root Cause Analysis" or "Root Cause Failure Analysis" have become quite popular over the last several years. There are a number of consultants offering a variety of techniques to perform root cause analysis. True root cause failure analysis can lead to dramatic improvements in reliability and overall performance. In addition, when individuals embrace many of the concepts involved in real root cause analysis, it can lead to significant insights concerning how their organization really functions.

Root cause analysis applies to a broad range of problems but for purposes of this discussion we are concerned only with identifying the root cause leading to the failure of physical systems or equipment. If you have been following the path described in past chapters, you have:

1. Started with a Malfunction Report

2.   Continued using available information to perform a diagnosis
3.   Used the recommendations coming from the diagnosis to perform troubleshooting
4.   Used troubleshooting findings to identify required repairs
5.   Classified the identified defect as a specific Failure Mode
6.   Performed Failure Analysis to identify the Failure Mechanism

In each event you have been focusing on a piece of physical equipment or a component with a defect.

In root cause analysis, your first step is to identify the "physical cause". Specifically, we are looking for a cause that is responsible for the identified Failure Mechanism.

For instance, if the failure mechanism is fatigue, and the fatigue resulted from unrestrained vibration or other motion, the physical cause might be a support or restraint that was not installed. In another instance, if the failure mechanism is corrosion, the physical cause might be the absence of protective coating. Further, the protective coating might have been worn away by abrasion, or scuffing by a loose or unsupported component.

It is important to acknowledge that no component ever designed itself, operated itself, inspected itself, or maintained itself. As a result, there had to be a human involved in creating the conditions for the failure. This level of cause is the "human cause". It is difficult but critical to identify it.

Cause – Fault – Blame:

These three terms need to be defined together because, all-too-frequently, people use them interchangeably.

Cause is the reason something happened. It is the first step in cause and effect. An individual may be the cause, but the failure may not be his fault.

Fault is the term used to describe the negative aspect of cause. If the effect was viewed as being good, the cause is viewed as being positive. If the cause results in an effect that is viewed as being bad, the cause is viewed as a fault. Again, it is possible to be at fault but still not be to blame.

The third term is blame. Blame has the most negative connotation because it implies guilt, reproach, and punishment.

When the search for cause degrades into looking for someone to blame, people become defensive, and useful information stops flowing.

When a failure occurs, there are negative consequences. It is a "bad thing". As a result, being identified as the "human cause" is also viewed as being "bad". Here it is important to distinguish between "cause" and "fault". Something I did might have been the cause, but that does not mean that I am at fault or that I am "bad".

Most people just want to perform their job as well as possible. There are very few instances where people are willfully negligent, or make mistakes on purpose. In these few cases, it is best to separate these individuals from the workforce, where they are a source of contamination and ill-will. But these instances are a minority.

If you are able to identify the human cause without fixing "blame" you will also be able to identify the latent or systemic cause. As stated earlier, since effectively no one causes a failure on purpose, there is always another cause, and you can find that cause in your system or organization. Taking a positive, non-fault-finding approach will create an environment where individuals feel comfortable in admitting to areas where there are lapses in knowledge or understanding, situations in which they were rushed, or where their job was not receiving their full attention.

If there is a weakness in knowledge or understanding, it could be lack of training. It may be that no training exists on a specific issue. It may be that training was generally ineffective. Or it may be that training was ineffective for only one person or a small group.

If an individual was distracted, a number of reasons are

possible. It might have been the end of a shift. The individual may have been dealing with personal issues. The work place may incorporate too many distractions.

In any event, there is always a latent or systemic cause that only management can address. Frequently, addressing a systemic cause will improve not only the problem being addressed but a variety of other problems that are being created by the same cause.

There is a saying that "your system is really perfectly designed to deliver the results you are getting". Unfortunately even those most devout devotees to root cause analysis seldom track cause all the way back to the systemic root. Identifying systemic causes are frequently threatening to individuals in authority. As a result, politically astute individuals typically avoid the conflicts and the risks associated with identifying the real reason(s) why things are going wrong.

For those who have the courage to "tell it like it is" and for those organizations that have the openness to listen, dramatic improvements are possible.

CHAPTER 11
BUCKETING INFORMATION

There is nothing new except what has been forgotten.

Marie Antoinette

Independent of how much data you collect and independent of the form you use to collect it there will always be a question that cannot be answered. Take it for granted that five, ten, or twenty years in the future you will be saying, "If I had only collected that one piece of information;" or "If only I had collected the information in a format that could have been more easily sorted". This is the result of finding new elements that are part of patterns not currently recognized for their importance.

An example of such an element is contaminants in oil. The contaminants have always been there and we have had the tools needed to measure them for quite some time. Despite that facility, it is only recently that many companies have begun to apply sophisticated oil analysis techniques to condemn oil. Most oil is still changed based on time or mileage and not the condition of the oil.

It is impossible to avoid the regret associated with not having collected all the data, but it is possible to minimize the regret by thinking of all the ways you may use the information, and what data will answer reasonable questions now and in the future. Here are a few of the questions you may be asked:

1. Is this a weather-related failure? If so, am I tracking weather conditions at the time of failure?
2. Is this kind of failure limited to a specific manufacturer, model, or age?
3. If this manufacturer is experiencing this kind of failure, what about other manufacturers of this same kind of equipment?
4. What was the specific age of the device at the time of failure?
5. How many KWH or miles did this device have on it when it failed? At what rate or percent of full load did this device operate?
6. Who was the operator when it failed?
7. Who was the last person who inspected or maintained it before it failed?
8. What was the failure mode?
9. What was the failure mechanism?
10. What was the cause of the failure?
11. How much did it cost to repair?
12. How long did it take to repair?
13. What was the history leading up to the failure?
14. Where was this located?

Anyone who has tried to use data in a high volume environment (lots of things needing to be analyzed) is aware of how useless are long narrative descriptions. It is impossible to extract useful data when individuals are allowed to input free-form information. Frequently this form of input is used to divert blame rather than document facts. As a result, it is better to identify all reasonable alternatives, then force individuals to choose from a look-up table. In addition, it is best to create mandatory fields that must be filled to complete the transaction and proceed to the next step.

For instance, if weather can be a factor, there probably are only a few abnormal weather conditions you are interested in seeing:

Instead of:

"It is a real blue norther' today."

You might want individuals to select from:

- Dry and Moderate Temperature (< 90-degrees)
- Dry and Hot (> 90-degrees)
- Dry and Below Freezing
- Wet and Moderate Temperature
- Wet and Hot
- Wet and Below Freezing

Here are a number of pieces of data you may wish to collect with each failure:

Equipment Number
Equipment Type
System
Location
Component (if known)

Serial Number
Lot Number (if known)
Design Data
Service Application (Name of stream)
Service Type (Acid, HC, Air, etc.)
History to Failure
Time of Failure
Age at Failure
Time in Service since Last Repair
KWH or Mileage at Failure
Who Reported Failure?
Who was Operating when Failure Occurred?
Who was Last to Maintain I Item Prior to Failure?
Weather Conditions at Time of Failure
Cost of Repair (Labor/Materials)
Components Replaced when Repaired
Time Required to Repair
Failure Mode
Failure Mechanism
Failure Condition
Physical Cause
Human Cause
Systemic Cause

By using the term "bucketing data", I am suggesting that almost all forms of data falls within a specific range in a continuous spectrum. If each item is allowed to occupy its own unique space, rather than falling into a "bucket" that represents a significant portion of the population, the data will be difficult to interpret. For instance, specific components may have failed at the following ages:

- 1 day
- 15 days
- 30 days
- 1 year
- 13 months
- 15 months
- 22 months
- 23 months
- 48 months
- 49 months
- 50 months
- 61 months
- 62 months
- 63 months
- 65 months

These specific ages are useful in performing a Weibull analysis. If you wanted to "bucket" the data, you might report

1. Three infantile failures
2. Five additional failures at less than two years
3. Seven failures at more than four years
4. Four failures at more than five years

From the scatter throughout the buckets, you can almost guess at three failure modes:

- One infantile failure mode
- One early failure mode
- One wear out failure mode

The infantile mode may be associated with poor construction or assembly. The early failure mode may be associated with a limited quality spill. The wear out mode may be normal end of life after having survived the other two failure modes.

Another example involves variations within a specific model. In past years, one locomotive manufacturer built three locomotives with XB70 (name changed to protect the innocent) as part of the model nomenclature. The first was the XB70N Tier 0 (falling under Tier 0 environmental rules). The second was XB70N Tier 1 (falling under Tier 1 environmental rules). The third was the XB70NAC (an AC locomotive falling under environmental Tier 2 rules). Both the XB70N locomotives were simplified locomotives with DC drive systems. The main difference between them was that the XB70N Tier 1 had a fuel injected engine and control systems contributing to enhanced environmental performance. The XB70NAC was unlike any previous XB70 locomotives and used the designation (in part) to benefit from the good reputation of the XB70N locomotives.

In this example, many systems within the XB70N Tier 0 and XB70N Tier 1 could be bucketed together. The fuel systems, and any systems with reliability impacted by the fuel system, would need to be bucketed separately. The XB70NAC would need to have all elements bucketed separately from earlier locomotives.

CHAPTER 12
ANALYSIS

Science is organized knowledge.
Wisdom is organized life.
                    Immanuel Kant

Analysis is defined as the separation of a whole into its parts for study and interpretation. Clearly there are few things in life that have such widely diverse interpretations as does the concept of an "adequate analysis".

When you are trapped in a traffic jam caused by the police collecting evidence associated with an injury accident, the amount of time needed to conduct an "adequate analysis" becomes increasingly less as the outdoor temperature climbs. When doctors are collecting information to determine how to handle the severe illness of a loved one, "adequate analysis" should take whatever time is required to deliver the results you hope for.

In major explosions and fires in industrial facilities, the regulatory bodies understand that most companies want to clear things away too quickly and get back to business as usual. As a result, in particularly serious situations, the regulators take control of the sites and do not allow the evidence to be removed until all the information needed for a thorough analysis has been preserved.

So, what needs to be said concerning "adequate analysis"?

First, it is important that a basic corporate value be "learning is important". It is doubtful if any executive in this era would argue with that statement, but there is more to the importance of learning than a simple platitude. There needs to be a belief that learning has value, and that ultimately learning will lead to progress. There is then value in making investment in learning. And sometimes the investment takes the form of inconvenience.

What source of learning can be more important than information concerning our own failures? Every day, companies spend scarce resources to send employees to attend courses leading to MBA degrees that the individuals may or may not use to the benefit of the company. On the other hand, these same companies are frequently unwilling to spend the money (or sacrifice the profit), to allow the time for evidence to be gathered that will help prevent similar failures in the future.

It is critical that companies begin with the attitude that they simply have to know the causes of failures.

Say that we have passed the point of convincing senior managers that learning is important and, despite the costs, they have decided thet they need to know the causes of failures. Where should we focus our attention?

Much like a detective trying to solve a crime, the focus is on collecting evidence. If the site of an equipment failure is viewed in the same manner as a crime scene, what evidence should be preserved?

- The failed component itself should be preserved. The anatomy of a failure provides a lot of clues concerning the cause.

- The conditions of associated systems (pressure, temperature, liquid levels, etc.) at the time of failure, and leading up to the failure may provide important clues.

- The actions of the operator at the time of the failure should be recorded as soon as possible, before time begins to change memories. What did the operator see, hear, or otherwise sense?

- Capture details of lubricant levels and conditions.

- What was the maintenance status? Was all PM current or were some tasks overdue?

- Do records suggest that past repairs were done correctly or were short cuts taken?

The most important point is that the information needs to be collected as quickly after the failure as possible for three reasons. First, if one waits too long, older failures lose importance compared to new failures. Second, as time passes, evidence pointing directly to the cause tends to get lost. Third, as time passes, people begin to "remember" things in a manner that distances themselves from any possible connection to fault.

An important attribute of a learning organization is its systems for capture and storage of data. You may believe that you currently have a great deal of data available to analyze failures. On the other hand, once you begin to analyze failures in earnest, you may find that both the data and the system used for storage are woefully inadequate.

Let's discuss an example. Say you are employed at a process plant that always has several pumps failing. Assume that two or three pumps fail every day. Also say, they have enough redundancy and have gotten good enough at performing repairs, that this stream of failures and repairs does not present a real difficulty. The reliability and availability of the plant is not being affected, and the plant maintenance budget has been set to cover all the costs. The main thing done by the engineers responsible for the pumps is to monitor the mean time between failures. As long as the MTBF trend is flat to slightly improving, everyone is happy.

Now let's assume that the world changes a little, as it so often does. Any of a number of changes could result in a need to manage these pump failures differently. Your budget could be cut. The plant throughput could be increased, or the feed stock could change. Operating conditions might change. Any of a number of things might increase the failure rate and decrease the MTBF.

In one instance, you may have to answer the question, "What do we need to do to improve things from the current situation?" In the second, you may have to answer the question, "What do we need to do to keep things from getting worse and restore performance?" In yet another situation, you may have to answer the question, "How do we do more with less?" In any event, you will need to understand the key elements and levers that are producing the current performance.

Mean Time Between Failures (MTBF):
MTBF is the arithmetical mean of the time between a number of failures of a single system or device, or all failures of a population of similar systems or devices. There are a variety of ways people choose to calculate MTBF and many are driven by the availability of information. Often, the MTBF is the time between successive failures. For some people, the MTBF is the time from the end of the last repair to the next failure. For others, the time in operation can be calculated accurately. The best approach is the one that takes into account the time the device is exposed to all forms of deterioration that will ultimately contribute to the next failure.

Performance is most often the product of a number of factors: the inherent reliability of the item as designed and manufactured, how it is operated, how it is maintained, how it is repaired or overhauled, and how it is modified. To improve performance, you will need to improve one or more of those factors. To restore performance, you will need to know which of those factors has changed negatively. To do more with less, you will need to understand which "levers" are important and which introduce added cost with little real value.

If you do not have the information, or it is not recorded and stored in a manner that easily shows trends, it will be impossible to tell what has changed or what needs to be changed.

A final point to be made concerning analysis is that the results of analysis need to be directly connected to action. Analysis should be designed to yield results that indicate if things are moving in the right direction or if they are moving in the wrong direction. If they are moving in the wrong direction, the reports from the analysis should stimulate corrective action. All too often, analysis is completed for the sake of analysis not for the sake of control and improvement.

Just for the heck of it, sometime, try intentionally making a mistake and report a result that should raise some eyebrows and cause some changes, just to see what happens. If nothing happens, the report is valueless. Try to find out what information will cause people to act.

CHAPTER 13
CREATING A COMPREHENSIVE
RELIABILITY PROGRAM

Knowing is not enough; we must apply!
Epictetus

In my handy pocket dictionary, the word *comprehensive* is defined as *totally inclusive*. For purposes of this discussion, I would like to define *comprehensive* in a different manner. A comprehensive reliability program is one that provides information and analysis at each phase during the lifecycle of your facility or equipment.

Lifecycle Cost:
   Lifecycle cost is the sum of all expenses associated with a system or device, beginning with the initial concept and ending with retirement. Lifecycle costs include the initial cost of procurement, operation, and maintenance. Costs of lost opportunity due to unreliability and outages are also included, together with the costs associated with unsafe conditions, environmental events, and expenses involved with addressing those issues, both during the life of the system or device and after its retirement if they continue afterward.

This chapter is dedicated to a brief discussion of the reliability techniques used during each of the numerous phases in a typical lifecycle.

Project Development Phase: Concurrent Engineering / Design for Reliability

In a role as a Reliability Director for a large chemical company, there were a number of occasions when I approached the Project Director to recommend performing reliability analysis as a part of the design process. In this situation, I could expect one of two responses. The first was, "You are too early. We haven't gotten to that point in the design yet." The second was, "You are too late. We have already completed the design and selected the equipment." Unfortunately, the amount of time between being too early or being too late was approximately three minutes and that occurred during the time I was in the restroom.

If a company wants to improve the inherent reliability of the units or systems they purchase or build, they must address reliability concurrently with other design considerations. They must include specific analytical steps that:

1. Test the expected reliability of the configuration and chosen components.
2. Compare expected reliability with performance requirements (e.g. the assumed uptime and reliability resulting in the expected annual production rate.)
3. Compare anticipated performance to performance after "tweaking" configuration, or component selection. (Compare lifecycle costs of several alternatives.)

A variety of tools is available to perform the analysis but it must start with establishing the requirement for Design-For-Reliability (DFR) to be completed concurrently with the physical design. If this expectation is not clearly established, the folks with the "take no prisoners" project management mentality will simply build the facility that produces the desired output, on-budget, and on-schedule. Those objectives are very important, but it is also important to have a highly- reliable and highly-available product. Without paying attention to reliability during the design, you only have a right to expect the performance resulting from characteristics that come from conventional design practices.

Configuration:
  In reliability analysis, several elements of configuration play an important role. One is redundancy of components in critical services, or redundancy of components with marginal inherent reliability. Another is series and close-coupled series configuration. For instance, if a process plant has two critical systems close-coupled in series, the failure of one will result in the immediate outage of the other. If some form of intermediate storage existed, a time-delay would be provided that would allow either for a re-start of the failed item, or a "soft landing" of the one still operating.

Project Design Phase: Reliability Block Diagram (RBD) Technique

RBD is one of the tools that should be used (but frequently is not) during the design of a complex system or equipment item. RBD uses a simple block diagram to represent all the elements and the configuration of the physical system. It is possible to assign reliability values to each element, and then use simple mathematical relationships to calculate the reliability of a system. Or you can use statistics to simulate the expected system reliability over the life of the system.

When RBD is used as a part of the design process, reliability and lifecycle characteristics used in the analysis are either taken from manufacturers' data or from information derived from experience with similar components. This analysis is used in several ways. Manufacturers use it to balance the initial cost of components with the value or the risk of failure over the warranty period. The lowest-cost component that will survive the warranty period, while delivering guaranteed reliability will be used.

Increasingly, well-informed owners are using RBD during the design process to make prudent choices in component selection and configuration. The OEM may offer a "basic design" that minimally meets reliability guarantees and balances first cost with future risks, but the owner may choose to upgrade specific components. The owners' lifecycle analysis is different from that of the OEM in several ways. First, the owner must balance first cost versus future risks for the entire life of the system, not just for the warranty period. Second, the owners' lifecycle cost needs to reflect the value or lost use of the asset, or lost profitability, in addition to warranty costs.

As a result, if the owner knows what reliability he has a "right to expect" from the "basic design" using RBD analysis, he may choose to select various upgrades in either component choice or configuration.

RBD finds application in analyzing and understanding existing systems for a variety of reasons. The most common reasons are:

1 The owner wants to increase reliability and needs to know the most effective place to invest limited resources.
2. The owner doesn't really know "what he has a right to expect" and RBD is a useful way to reconcile the owner's expectations with reality.

In the first instance, a RBD model is developed for the current configuration using the reliability of the current components. After the anticipated reliability of the current system is calculated (or simulated), the configuration or component selection is modified one piece at a time. For instance, redundancy might be modeled by adding a spare pump or circuit to a critical application. Then the model is re-run. (Although the absolute value of modeled reliability might seem different from "real life", the modeled difference between pre-modified and post-modified will provide a realistic basis for comparison of alternatives.) In another example, an individual element that has marginal reliability might be replaced with another, more robust component. After several possible upgrades are modeled, the changes providing the needed improvement at the lowest cost are chosen, and modeled in combination.

RBD provides an excellent tool for both identifying the best opportunities and for determining what you realistically "have a right to expect". Once data is gathered using the information collection system described in this book, the results of a RBD analysis based on that data will become increasingly accurate. Your own data reflects how well you operate your equipment and how well you maintain your equipment. The data provided by OEMs, and in reliability tables, tends to reflect how the equipment or components perform for the "average" owner. You may take better care of your equipment, or you may expose your equipment to harsher service. In either event, the data coming from your own equipment is likely to provide you with a much more accurate estimate of how improvements will perform.

Project Design Phase: Engineering

A requirement to conduct Design for Reliability will highlight the need to emphasize the application of reliability techniques at the same time as the basic engineering design, but the importance of sound basic engineering cannot be over-emphasized. All too often, the design engineers do the basics but ignore unusual issues found in the environment where the system is expected to function.

For instance, the design engineer should consider extremes. Frequently, there are situations that are unusual but should be expected. Systems need to be able to survive extremes in weather: maximum summer temperatures, minimum winter temperatures, very wet conditions, high winds and blowing moisture. Although these conditions are not everyday occurrences, they are also not unexpected conditions.

Other environmental characteristics are vibrations, dust and dirt, exposure to UV radiation, exposure to chemicals, high g-forces, and other forms of external loads.

In an unusual example where g-forces were not adequately considered in design, a manufacturer of mobile equipment found that g-forces resulting from starting and stopping caused failures. In this instance, several critical contactors had always been mounted on side walls (where movement of the contactor was at a 90-degree angle to the normal direction of motion). In a design change, the contactors were moved to the back wall of the electrical cabinet (where opening and closing motion was parallel to the normal movement of the vehicle). It was then found that the strength of the holding coils in the contactors was inadequate to sustain contact during starts and stops, so intermittent breaks occurred leading to failures.

As described elsewhere, the failure mechanisms for mechanical systems are corrosion, erosion, fatigue, and overload. The basic design engineering should address the following questions:

- Is it possible that metals with differing potentials will be exposed to an electrolyte that will set up a corrosion cell?

- Are there sources of abrasive materials and is there an energy source to move those materials resulting in abrasion between moving parts?

- Will normal movements or vibration result in alternating compression and tension? If so, is the loading above the fatigue limit? How many cycles will result in a failure? Will any components be changed before failure occurs?

- Is the basic design of the selected component capable of handling all loadings? What is the source of maximum loading? What conditions could result in loading greater than the design maximum? What is the likelihood of that over-loading? If the likelihood is greater than 1% over a 30-year life, the impact should be considered. What will result?

The same kinds of questions concerning failure mechanisms for electrical systems should be addressed during the design:

- Are components sensitive to supply transients? What happens if there are spikes in voltage or current? What will happen if the frequency changes?

- Is there a possibility the system will experience a stall? If so will it overheat? What damage will result?

- Will this system experience the electrical equivalent of fatigue? Will there be regular situations where the system operates at conditions greater than design limits but less than is needed to result in an instantaneous failure? What will be the cumulative effect of such events?

- Is all insulation designed to endure the maximum temperatures possible?

- Will the system be exposed to any unusual chemicals? If so, have materials been selected that will withstand those chemicals?

- Will non-metallic components be exposed to direct sunlight? Have those components been selected to withstand UV radiation?

- Will the environment contain materials that can abrade contacts or sensitive surfaces?

- Is the system exposed to vibration or g-forces that are likely to cause loosening, chattering, or deterioration?

This is not an exhaustive list of questions, but it provides a reasonable start on the set of questions that should be answered to ensure a robust design.

Project Execution Phase: Procurement

Increasingly, companies are preparing less definitive specifications and depending more on "industry standards". This approach has the benefit of being more cost effective, but carries some risks.

Improved cost effectiveness results from not specifying characteristics that are unusual and inconsistent with what most people want. On the other hand, there are issues that tend to differentiate one customer from another, and vanilla specifications do not protect the differentiation.

An example of this is a simple mechanical seal. Many different kinds of water pumps share a common type of mechanical seal, and a failure of the mechanical seal is the most likely reason the pump will fail. Some of those seals end up in automobiles, some end up in boiler feed pumps supplying steam boilers driving turbines in electrical plants, and some end up in locomotive cooling systems. In one instance, an individual's comfort and convenience depends on the pump and seal. In another, the supply of electrical utilities to a city might depend on the pump and seal. In the third, the movement of critical supplies from one  point to another might depend on the pump and seal.

Specifications used to procure the encompassing system (automobile, boiler feed water pump or locomotive) can vary widely with respect to the amount of details provided in the seal specifications. There are no specifications used to purchase a typical automobile, much less the water pump seal it contains. The boiler feed water pump specifications concerning the mechanical seal are likely to be fairly detailed, describing materials in wearing faces, elastomers, etc. On the other hand, although the locomotive reliability is business-critical, it is not common practice to specify the water pump seal.

Generally, the typical person purchasing an automobile would not know how to begin to assemble a general specification. The typical automobile buyer specifies one of a dozen exterior colors and one of three to five interior trims. In addition, he or she specifies a variety of options, like engine size, suspension options, and sound system. Effectively all automobile buyers depend on accepted standards, and the influence of regulators to ensure basic components are properly designed and specified. They do not specify the size of the brake shoes or the number of bolts holding the wheels in place. They assume that these critical factors are properly engineered by the manufacturer and if not, the regulators will step in. Generally this trust is well-placed and the major car makers produce safe vehicles.

On the other hand, think for a moment about automobile reliability. The typical automobile buyer treats reliability much like safety. He or she trusts the manufacturer.

Now think about the vehicles you have owned. How long do they last for you? How long do you drive a car and why do you get rid of it? Think about the automobile you had two cars in the past or maybe three. Do you see any of that version on the streets any more, or have they all disappeared?

In recent years, I purchased two new vehicles with which I experienced completely different reliability.

The first example was a 1989 Oldsmobile Cutlass Calais with a Quad Four engine. It was a red two-door with a five-speed manual transmission. When new, this was a nice little car. It was well powered and got reasonable mileage. After a few years, I had a fairly long commute to work and I no longer felt it was reliable enough to get back and forth each day, so it was passed down to my son. Any parent with a teenage son knows the rest of the story. It lasted another year or so and then had to be replaced. The point of the story is that within ten years of manufacture, you seldom saw any of these cars on the road. They were all gone.

Quad-Four Engine:
An engine with four cylinders and four valves per cylinder. Cars with this engine could be recognized from the rear by the white smoke. The white smoke resulted from the coolant that the flexible cylinder heads allowed to pass into the cylinders and be burnt with combustion gases.

The second example was a 1994 Ford F-150 Pickup Truck with a 300-cubic inch, six-cylinder engine. I drove this truck for  eleven years. I drove it trips as long as four hundred miles before I sold it. I traded it because I needed a four-wheel drive with more passenger space, not because of reliability. Again, the point here is that you commonly see this model of truck as well as trucks much older, being used every day on the street. They provide good reliable transportation for a long life.

With both vehicles I trusted the manufacturer in the same manner concerning the reliability of the product. I was not able to provide more or different specifications for the Ford than for the Oldsmobile, but the Ford's reliability far outperformed the Oldsmobile. Getting back to the typical question, "What did I have a right to expect?" Based on what was specified and purchased, I had a right to expect the performance I received from the Oldsmobile. I was fortunate to receive the performance I received from the Ford.

Now let's take our discussions back to the boiler feedwater pump in the power plant and the locomotive. What does the specifier of each of these examples have a right to expect?

Clearly, the individuals in the power plant have had experiences that have taught them to specify equipment to a much greater level of detail. Detailed specifications are the "industry standard" in the utility industry. On the other hand, the railroad industry over the last 150 years has developed into an industry where the locomotive specifications are a blend between the limited specifications used to purchase road vehicles and the somewhat more detailed specifications used to identify user-specific requirements.

Utilities can expect the level of performance their knowledge and experience has earned for them . . . . but no more. By providing detailed and exacting specifications, the owner is assuming responsibility for results. Railroads can expect the level of performance that their relationship with their supplier has earned for them ..... but they legally have a right to demand very little. They trust the OEM to make prudent choices, and to stand behind their product when the choices are poor. The OEM then takes on the risk of failure. If the OEM is to be expected to stand behind their product, they need to be able to pass along the cost of the risk with the price of the product.

Specifications covering issues important to reliability end up being a balance between what you know and the trust level you have developed with your suppliers. Increasingly fewer and fewer companies maintain the staff needed to design and engineer all the equipment they use, so they must depend upon the suppliers to support their product. Purchasing departments who wring the last dime of profit out of every purchase create relationships in which everyone has to look out for themselves. In the long term, that just doesn't work.

Project Execution Phase: Construction/Assembly

With frequent workforce reductions in the last few years, many of the positions that are viewed as "non-core" or "non-essential" have been eliminated. Many reductions occur at a time when capital spending is limited, so individuals involved in project development or construction are viewed as expendable. As  a result, many of these positions have been eliminated by down-sizing.

As the design, construction and QA/QC personnel are eliminated, their tasks are turned over to "others". Frequently, after a few years, companies find that the "others" they trusted to handle these non-core issues are not motivated by the same things as the companies they represent. The business model of an engineering/construction company is different from that of the owner. Engineering/construction companies make their money based on strict control of scope, schedule, and budget. These factors are all measured at the conclusion of a construction project. Reliability, availability, and maintainability can only be measured much later, after individuals involved with the project are long gone.

One might believe all the issues that determine inherent reliability are decided during design, engineering, and preparation of specifications. That would be true if all assembly or construction were done equally. But it is not.

As an example, let's consider automobile manufacturing. Years ago, regulatory bodies recognized that a small portion of the vehicles produced each year did not meet the same reliability performance of the overall population. As a result, they passed "lemon laws" to protect the individuals who purchased those vehicles. Studies have found that the "lemons" were frequently manufactured on Mondays or Fridays. The normal QA/QC practices followed through the rest of the week were not adequate to identify the defects being built into the lemons.

This same kind of problem exists in other industries. The work done on weekends and off-shifts can produce "lemons" in any situation where quality is not properly monitored.

There is an old saying that you should "expect what you inspect". Although it is frequently impossible to inspect everything, it is important to determine what is most important to your business success and to inspect those areas. The personnel performing inspection should be responsible to the owner rather than the builder.

Asset Operation Phase: Start Up

Much like construction or assembly, the start-up or initial commissioning and use of a system can have a lot to do with its ultimate reliability. I will use an extreme case as an example. A number of years ago, I was working in a refinery where we performed some heat-exchanger bundle cleaning almost every day. Large twenty to twenty-four foot long tube bundles were removed from shells using two hydraulic cranes to support the bundle as it was pulled out by means of a tractor. During this time a manufacturer developed a single piece of equipment that could do the job of the two cranes and the tractor. Besides being more efficient, the method was safer.

The hydraulic bundle extractor arrived and appeared to be an answer to a maiden's prayer. It was designed well and worked well. However, after some time we began having problems that caused delays. We experienced frequent hydraulic leaks. After putting up with poor service for several weeks, we began trying to understand the problem. We were finding small metal shavings in hydraulic seal components. After some searching, we determined that during construction of the large hydraulic sump that was used as the base and frame for the vehicle, the fabricator had allowed metal shavings and coupons from hole-saws to fall into the sump. The sumps were never cleaned out before start up and delivery of the equipment to the customer. When the equipment arrived, we thought it was ready to use, so we never "commissioned" the equipment. Had we known about the poor manufacturing processes, we would have cleaned out the sump and flushed all the hydraulic lines. As it turned out, all the hydraulic seal components had to be replaced. By this time, the equipment left a bad taste in our mouth so we returned it to the manufacturer.

Commissioning and start-up is a critical part of setting equipment on the right path to a long and reliable life, or it can be a time when unnecessary defects are introduced that can doom the system to a life of being viewed as "never was any good."

Asset Operations Phase: Operations

Over the last fifteen years or so, several companies and plants have introduced an organizational approach known as Total Productive Maintenance (TPM). In TPM plants, the operator is expected to perform a variety of the maintenance tasks that previously were assigned to craftspeople. It is impossible to adequately describe all the elements of TPM here, but it is important to highlight the fact that equipment seems to operate better when the people who operate it also maintain it, and know what actions are likely to cause damage.

In attempting to describe the psychology of the man-machine interface, there is a human characteristic known as the "device mentality". My best description of the "device mentality" is drivers who only know about the accelerator, the brake, and the steering wheel. They have no idea how any part of their car functions, other than the response to those three things. Cars belonging to people with this "device mentality" have a short and unreliable life. Their owners tend to ride the brakes. They tend to accelerate too quickly. They tend to under-maintain their vehicles. The list goes on and on.

When people understand how things work, they also begin to understand how certain actions may cause additional wear and tear, and they begin to avoid those actions. When people know the behaviors, sights, and sounds of a problem when it is just getting started, they can shut equipment down before major damage occurs.

For example, changing oil that has become discolored

gives a pump a whole new lease on life. Failing to change dark oil may lead to a bearing failure that can otherwise result in additional damage. Even after a bearing is renewed, there can be "infant mortality," or chronic problems associated with not getting the equipment quite right. The overall effect of allowing a problem to get started is impossible to quantify. The best policy is to employ personnel who will prevent defects before they can cause failures.

One of the best descriptions of TPM is in a story I heard a long time ago. An older gentleman was responsible for maintaining all the big reciprocating compressors used for gathering gas out in the Texas oil fields. He had been asked to give a talk to a number of other engineers who had jobs similar to his, but who had never been nearly as successful in providing good, reliable machines. He described a detailed process where he obtained a special kind of wood and had the wood made into special "listening sticks" for the equipment operators. The sticks had to be a special length and all the dimensions and finish had to be carefully controlled. Next, he had small bulls-eye target stickers made and installed on all the compressors at points that roughly corresponded to each of the bearings on both the driven and the driving machines. Finally, he had each of the operators trained to hold the stick up to the area immediately ahead of their ear and consecutively place the other end of the stick on each of the bulls-eyes during each round (several times each day).

About this time, someone in the group listening to the talk spoke up and asked, "How did it work? Were the operators able to hear vibrations, or the sounds of defects that were forming in the bearings?" The old man responded, "No, when they got that close, they could see if anything was wrong."

Asset Operation Phase: Inspection

As systems and equipment age, deterioration due to wear and tear is normal. Considering mechanical systems only, three of the four failure mechanisms (corrosion, erosion, and fatigue) are forms of deterioration that occur over time. Each of these deterioration mechanisms is a fact of life, and apart from "gold plating" everything, they must be dealt with as systems age. The challenge is to intervene before deterioration leads to a failure, but after most of the useful life has been consumed.

 Most forms of deterioration are "ratable" meaning that they occur at some defined rate. Take uniform corrosion for example. Piping and other fluid-handling systems will deteriorate at a relatively fixed corrosion rate. You can determine this rate by measuring the thickness of the metal being corroded at regular intervals and calculating the rate (in mils per year) that metal is being removed during each period. To intervene before failure you have to: 1) understand the corrosion rate and 2) understand the minimum metal thickness needed to prevent failure. By calculating the time at which the corrosion allowance has been eliminated, you can identify the latest time you can intervene before failure.

Uniform corrosion is one of the more easily quantified forms of deterioration, but many forms of deterioration can be evaluated in some analogous manner. The pattern is inspection → evaluation → intervention. The objective of the inspection is to determine both current status and deterioration rate. Knowledge of current status, deterioration rate, and condition at the point of failure, leads to timely intervention.

It is possible to fill books on the subject of inspection, but the philosophy gets down to the simple steps of:

Inspection → Evaluation → Intervention

In my experience, inspection has been more closely associated with equipment integrity than reliability. People perform inspections to avoid catastrophic failures that can adversely affect safety. As a result of boiler failures, states created regulatory bodies and the regulatory bodies created the requirement for regular inspections of boilers. Over time, similar events happen that involve pressure vessels and a variety of other equipment items. These events can produce untoward results if not adequately monitored and maintained. Association of inspection with integrity does not minimize its importance with respect to reliability.

Intervention:
    An action that reduces deterioration or eliminates a defect before a failure can happen.

Preventing catastrophic events also improves reliability, but the scale of events to be prevented for improved reliability is typically much smaller than the catastrophic events targeted by "integrity-based" inspections. Reliability-based inspections are relatively minor in nature. Rather than preventing major explosions, reliability-based inspections are aimed at detecting and preventing the deterioration that leads to failures, rather than the failures themselves. The key word is prevention.

As a result of the significantly increased frequency and greatly decreased scale, reliability-based inspections are commonly done by someone other than the individuals who perform integrity-based inspections. Either operators or craftspeople who are continuously assigned to areas can conduct minor inspections during each "round," or several times per day. If someone is keeping his or her eyes and ears open, there is no reason for problems to progress to catastrophic proportions.

Asset Operation Phase: Maintenance

Maintenance can be performed in a wide variety of ways. Some of these ways support reliability, and others create defects that lead to poor reliability. Earlier, we discussed the difference between proactive maintenance (predictive and preventive maintenance), and reactive maintenance (after a failure has occurred). It is clear that performing an intervening task before a failure occurs is less expensive and results in fewer interruptions than waiting for failures to occur before acting. But there are a variety of ways in which reactive or repair maintenance can be done. Some of them are consistent with improving reliability and others are not.

At one extreme is the old baling wire and chewing gum approach. At the other extreme is the practice of trying to make everything better than new. I support neither extreme.

Baling wire and chewing gum repairs are the result of a "fire-fighting" approach to maintenance. You are never quite sure when the next fire will break out so you are always in a hurry to put out the last one. Rather than restoring inherent reliability by taking the equipment back to its original specifications, this approach simply fixes what is broken, up to the point that the unit can be placed back in service. In this event, the mean time to failure after a repair is frequently a fraction of the MTBF of a new (or properly maintained) unit.

The other extreme is to try to make everything "better than new" during repairs. I once worked in a refinery where the leader of the rotating-equipment discipline viewed every repair as a new re-engineering project. Rather than replacing worn shafts with a new OEM part, he would have them sprayed with a metal coating. Most often this would result in the bearings needing to be resized and frequently the seals also. On all future repairs, any original equipment parts were valueless. Components had to be fabricated from scratch. This routine resulted in longer outage periods and reduced availability. Equipment was out of service longer, so plant reliability and availability suffered.

The best way to perform maintenance is to begin by engineering things correctly, work out any bugs during startup and early operation, then maintain equipment using OEM parts and restoring original tolerance, fits, and clearances. Repair maintenance should restore inherent reliability each and every time without short cuts. Short cuts lead to more short cuts and sooner or later everything in your facility looks like a dog's breakfast.

Asset Operation Phase: Turnaround/Overhaul

Sooner or later almost every plant or equipment item will need to go through a planned outage for major maintenance. Some people call these turnarounds, some call them outages, some call them overhauls, and some call them shutdowns. Whatever you call them, the objective is to restore the equipment to a condition that can provide reliable service until the next planned outage.

I have been involved in outages lasting as little as a few days and costing tens of thousands of dollars to ones lasting several months and costing well in excess of a hundred million dollars. In any event, reliability, or preparing for a reliable run, is one of the most important aspects of any outage.

In applying reliability management concepts to an outage, the first step is to identify two critical premises. The first premise is the planned duration of the run between this outage and the next one. The second premise is the level of reliability that will be expected between this outage and the next one. These two premises will go a long way toward determining the scope-of-work, and the needed condition of the equipment at the conclusion of the outage (the start of the next run). All the items in the plant (or on the equipment being maintained) must be in a condition that will ensure that it survives until the next outage.

One of the sections in the next chapter discusses "precision maintenance" or the practice of:

1. Identifying "as-found" conditions at the start of any maintenance activity.

2. Based on the difference between "as-left" conditions (at the end of the previous repair) and "as-found" conditions (at the beginning of this repair), calculating the deterioration rate.

3. Identifying the required "as-left" conditions at the conclusion of the current maintenance needed to ensure that the item will survive for the desired run-length.

The same concept as described for "precision maintenance" must be applied to all parts of the plant (or equipment) during an outage to ensure the required reliability for the intended run-length.

Codes that describe requirements for management of uniform corrosion in piping and pressure vessels provide a good example of how precision maintenance concepts can be applied to "force" systems to be reliable for a specific period of time. In those systems, several critical factors are used to characterize the condition and integrity of the system:

>MW – Minimum Wall thickness, or the thickness of metal needed to safely retain the operating pressure.

>CA – Corrosion Allowance or the thickness of metal that can be corroded away before MW is reached.

>CR – Corrosion Rate or the rate (in mills per year) at which uniform corrosion is proceeding.

>Life – Life is the number or years that the system will survive based on the current corrosion allowance and corrosion rate.

>Half Life – Half Life is half of the current life.

Using this system, the metal thickness is measured using ultrasonics or other thickness measuring techniques at regular intervals. The thickness during the prior measurement is analogous to the "as-left" condition. The thickness during the current measurement is analogous to the "as-found" condition. The corrosion rate (or deterioration rate) is calculated as follows:

$$CR = \frac{\text{Prior Thickness (mils) minus Current Thickness (mils)}}{\text{Time between measurements (years)}}$$

The available life is then calculated using the following equation:

$$\text{Life} = \frac{CA \ \ (\text{mils})}{CR \ (\text{mils per year})}$$

In piping and pressure vessels containing toxic or hazardous materials, it is required that the next inspection occur at the "half-life". This timing ensures that half of the corrosion allowance is still intact at the next inspection. If some unforeseen event occurs that accelerates the deterioration rate, it is likely that equipment integrity will remain intact and any problem will be discovered before it results in a failure.

This same concept of "managing deterioration" can be applied to other forms of uniform deterioration and wear. The current condition and deterioration rate can be compared to the condition in which a component will no longer be serviceable. If that point occurs after the end of the desired run, repair or refurbishment is not needed. If the minimum serviceable condition will be surpassed some time during the desired run-length, it will be necessary to refurbish the component.

A successful outage (from the standpoint of reliability) is one in which all critical components are inspected, analyzed, and addressed in the manner described above. Equipment and components that can be repaired on-line without causing an outage are the exception. They should be repaired (using precision maintenance methodology) outside the duration of the outage.

Asset Operation Phase: Expansion/Modification

Expansions and modifications should be treated in the same manner as new equipment and facility designs. The Design for Reliability should be completed at the same time as basic design.

I recall one occasion when a young electrical engineer was taught the basic essence of RBD analysis as part of a brief reliability training program. At the time as this training, a major revision to the electrical distribution system in his plant was being prepared to support a plant expansion. It was not a practice of the engineering contractor to perform DFR or compare the future configuration to the existing configuration using RBD.

Instead, the young engineer performed a RBD analysis and was able to show that the new configuration of the proposed electrical system would result in a substantial loss in reliability from the current configuration. This result was unacceptable to plant management. When plant management showed the analysis corporate managers they were surprised to see the added risk of failure. Project managers had always described the changes as being "improvements". In this instance, RBD analysis of the revised configuration justified a change in the design and prevented future problems.

Increasing numbers of companies are making some form of Design for Reliability a mandatory part of project development. They are including reliability and availability hurdles with capacity, quality, and efficiency requirements for both green-field and modification projects.

Asset Operation: Continuation/Renewal

As a final comment for this chapter, let's discuss the general area of continuation or renewal. Every asset has an annual depreciation rate. Although many assets continue to operate long after depreciation is complete, few operate reliably without reinvestment at a rate close to the depreciation rate.

Let's take two examples that people seldom consider: structures and electronics.

Steel structures are often ignored when re-investment to offset deterioration is considered. It may not be necessary to re-invest at the tax write-off rate, but a significant level of re-investment is needed to ensure structural integrity and appearance.

It is often thought that electronic components become outdated long before they wear out, but that does not always happen. Depending on the environment, electronic boards can become overheated or dirty (causing shorts), or be removed and reinstalled numerous times (causing connections to wear out). It is not uncommon for the application to outlive the hardware and if it does, ongoing reliability depends on replacing elements that might not typically be viewed as a wearing component.

By highlighting these two extremes I hope to point out that you need to consider them and everything in between if you want to maintain inherent reliability over the entire useful life of the asset.

CHAPTER 14
GENERAL COMMENTS ON RELIABILITY METHODS

Science is what you know. Philosophy is what you don't know.

Bertrand Russell

I selected the above quote to introduce this chapter because it seems to sum up reliability analysis in many situations. For some people, reliability is a science and for others it is a philosophy. One can be very much dedicated to the philosophy of good reliability without doing the heavy lifting needed to produce high reliability.

The discipline needed to achieve high reliability is like a religion. A person can attend church every Sunday and still not be a religious person. Similarly, it is possible to attend to reliability issues every so often but not all the time. You must then expect mediocre results. To achieve the best results, you need to turn your reliability programs from a philosophy into a science and from a hobby into an avocation.

> **Discipline:**
> Performing according to an established set of protocols and procedures. Performing consistently. Performing to the same high standard all the time.

An issue for your consideration is that all the steps required to process a failure (Malfunction Report – Diagnostics – Troubleshooting – Failure Mode Identification) happen all the time. It doesn't matter if you have formalized the steps and trained someone to do them in an organized manner, or if they happen in the natural course of things. They happen. It is your choice if they happen in a structured manner that collects and analyzes critical information, or if you simply allow the process to flow guided by the "philosophy" that reliability is "nice to have".

For instance, in your "system" a failure might occur and a mechanic simply goes to work trying to correct it. He might not be the right person and he might not have access to the right tools or information, but he will still need to perform some form of diagnosis. He may be woefully inefficient. He may even "fix" the wrong problem several times before getting to the actual problem, but he will perform some diagnosis.

The same is true of each of the other steps.

My reason for highlighting this fact is that, when you start to implement a reliability program, detractors frequently say, "We don't need to do that. Our people know what to fix and how to fix it. Performing diagnosis would just cost more money and take more time." It is important to recognize that you are paying for diagnosis anyhow. There is no way around it. By formalizing the process, and organizing the data collection and record keeping, it is possible to make the activity much more efficient and effective.

Another issue highlighted in this text is that after you have employed this process for some time, you will have a lot of accurate information concerning the reliability of your equipment. By itself, using this process will lead to significant reliability improvements. But, once you have the information available, you will be able to go much further. This chapter is devoted to the reliability techniques that become available, once you have access to well-structured and accurate information.

Reliability Block Diagram (RBD) Technique

Many techniques are useful in evaluating the expected reliability of a system, but the Reliability Block Diagram technique (RBD) is one of the most useful. An RBD is drawn using simple blocks to model the elements of the system that can have an impact on reliability. Static components that have little or no effect on reliability are frequently ignored. There are two forms of RBD analysis. One uses simple mathematical expressions to calculate the anticipated system reliability. The other uses a number of life-cycle simulations to determine the expected reliability performance based on statistics.

Let's discuss the approach using simple mathematics. (A number of commercially available simulation models can be found on the internet and that resource is left to the reader to explore.)

The first step is to understand the most basic element., of the Reliability Block Diagram.Depending on the level of your analysis, an individual box can represent a small part, or a complete operating unit or plant. In either instance, the number in the block represents the expected reliability of the element that the box represents. The reliability of that element is calculated as follows:

$$R(t) = e ** - (t / mtbf)$$

Where,

R(t) is the reliability or likelihood of operating for the period of time t without a failure.

** represents "to the power of" the following exponent.

mtbf is the mean time between failure for the element represented by the box.

For instance if the mean time between failure is one year and the time interval of interest is one year then,

$$R(1) = e ** -1/1 = e ** -1 = 0.368 \text{ or } 36.8\%$$

Somewhat surprising is the fact that if the mean time between failure is one year, the mathematical likelihood that the device will make it through any one year period without a failure is less than 40%! Keep in mind that in a large population, half of the devices fail in a period less than the MTBF.

In this example, the element would be represented by
the block as follows:

The likelihood of failure of any element is one minus the
reliability.

Assume that we have a device that has an mtbf of ten
years. Then the likelihood of failure in any one year is
approximately 0.10% and the reliability is approximately
90%.

For simplicity in the following example, we will assume
that we are exclusively using elements with reliability of
90%. This assumption helps simplify the math and it
helps to portray more clearly, the impact of adding elements in series as compared with those in parallel (or
redundant) configurations.

For a series configuration, the reliability is the result of
simply multiplying the elements together. For example:

In this case the "system" reliability would be:

$$R = 0.9 \times 0.9 = 0.81 \text{ or } 81\%$$

It should be no big surprise that the series configuration
has a composite likelihood of failure almost twice that of

a single element.

Now let's consider two elements in a parallel or redundant configuration:

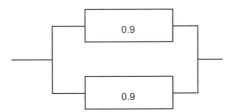

In this example, the "system" reliability is the result of the following equation:

$$R = A + B - (A \times B)$$

Where A and B represent the reliability of the two elements.

In this example, the reliability will be:

$$R = 0.9 + 0.9 - (0.9 \times 0.9) = 1.8 - 0.81 + 0.99 \text{ or } 99\%$$

In other words, the redundancy significantly reduces the likelihood of failure of the system.

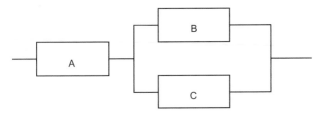

For the sake of completeness, we will analyze a system that includes elements in both series and parallel:
In this example the system reliability is calculated as follows:

$$R = A \times [B + C - (B \times C)]$$

where A, B and C are the reliabilities of the various components.

Two comments concerning the use of this RBD methodology:

1. The reliability resulting from the calculations may not be absolutely accurate, but the tool can be used to identify the blocks (or components) that are likely to have the greatest impact on performance. By substituting 1.0 for the reliability in each block sequentially, and calculating the resulting reliability, then by comparing the results you can identify the element that is causing the biggest problems. If you can replace or upgrade the performance of the element resulting in the largest improvement, you can have the greatest impact with the least expense.

2. The precise number resulting from the calculation might not be completely accurate, but a comparison of two numbers using the same technique and assumptions for both will provide a valid comparison. For instance, if you use the same techniques and information sources for two different configurations, the relative performance is likely to provide an accurate portrayal of how the two systems will compare.

"Static" components (like tanks, pressure vessels, and piping), are frequently ignored in the analysis, but there are occasions when it is useful to include them.

1. If a significant proportion of reliability failures results from the failure of these elements, they should be included in the analysis.

2. It is sometimes possible to improve the reliability of "close- coupled" facilities by providing intermediate storage capacity. Assume that a tank is provided between two plants that normally feed directly from one into the other. Also assume that this tank is kept half-full. Then, if the upstream plant fails, the down stream plant can temporarily draw feed from the tank. If the downstream tank fails, the upstream plant can continue to run intermediate product to the tank while the downstream plant is being repaired and re-started. Currently-available computer programs are capable of modeling the intermediate storage to determine optimum capacities and operating levels.

Reliability Centered Maintenance (RCM)

RCM is one of the forms of Failure Modes and Effects Analysis (FMEA) used to help ensure the reliability and integrity of complex systems. RCM comes in a number of shapes and sizes. It seems that individuals who are advocates of one form of RCM are dead-set opposed to the other forms.

> **Failure Modes and Effects Analysis (FMEA):**
>   Any of a number of structured analytical tech-
> niques that identify the failure modes possible
> within a system and link those failure modes to the
> resulting effects. The effects are characterized by
> their associated risk (impact times likelihood) and
> the cumulative risk for the system is calculated.
> The objective of the analysis is to identify actions
> that will manage failure modes, resulting in a
> reduction of the associated risk.

Depending on how the form of RCM being used is structured, it can do a few things, or it can incorporate a number of objectives into one analysis.

In the following discussion, I will try to briefly describe
RCM in a general manner. I am sure that this description
will not satisfy everyone, but it is a reasonable approach
and includes many of the alternatives that are possible.

• The analysis begins by clearly identifying a system, or
  a portion of the plant or asset, that will be analyzed. It
  is important to set clear boundaries because those
  boundaries are critical in identifying the included
  functions and in making assumptions.

    An example of a boundary issue is: where will
    specific isolation valves be analyzed, in this sys-

tem, or in an adjacent system? Critical equipment that is capable of producing failures must clearly be in the system most affected. Another example of a boundary issue is utilities and instrument signals. It is typically assumed that utilities will be supplied, and signals will be correct. If there are problems, they will be analyzed in their own systems.

• The next step is to identify the function or functions performed by the system being analyzed. This is an important step because the function(s) is the element you are trying to preserve. For instance, if the function of a boiler is to produce 100,000 pounds per hour of 100 psi steam, then it is clear that when steam production is halted, the intended function is not being achieved. But how about when the boiler is capable of producing only 90,000 pounds per hour of steam or only 90 psi steam? Is the function being achieved in those instances? It is important to clearly define the "specifications" for the required functions.

Assume also that the boiler makes a certain amount of noise while operating. Is the sound level a critical part of the function? Do you wish to identify proactive tasks that will preserve the sound level? If so, you must include this data as a part of the definition of the function. If not, don't bother with it.

• The next step is to identify all the forms of "Functional Failure" (FF). In the example above, if the specification of the function is to produce 100,000 pounds per hour of

100 psi steam at 90 dB or less there are several possible FFs:

1. Boiler is completely shut down.
2. Boiler is producing less than 100,000 pounds per hour of steam.
3. Boiler is producing steam at something less than 100 psi.
4. Boiler is operating at greater than 90 dB.

• The next step is one that depends on choices you have made concerning your form of RCM analysis. The choices involve economic analysis of FF. Most businesses that conduct analysis are interested in balancing the cost of prevention with the value of the failure being prevented. So it is important to identify the value of each form of FF. The total cost of a specific failure mode varies with the duration of the resulting event, so it is required that the cost of each FF be stated in dollars per hour or day.

In the example above let's assume that when :

1. The boiler is down, the cost impact is $24,000 per day.
2. The boiler is producing less than 100,000 pounds per hour (2,400,000 pounds/day), the loss is figured on a sliding scale at $0.01 per pound down to 60,000 pounds per hour, below which the boiler cannot operate.
3. The steam boiler is producing steam at less than 100 psi, let's assume that the receiving system operates at 100 psi, so the steam produced will not make it past the check valve into the system. This

condition is paramount to the boiler being shut-down.

4. The boiler operates at greater than 90 dB for any portion of a day, there is a fine of $10,000 per day.

- The next step is to "import" all the equipment and components that can result in any FF when the item fails. Again, there are several choices at this step. Some forms of analysis look at "dynamic" equipment only. Dynamic includes pumps, compressors, instrumentation, valves ….. things that move or regularly change states. This approach would exclude "static" elements like piping and pressure-retaining equipment. The choice of what is to be included should be based on which components regularly contribute to the "unreliability".

   Another choice has to do with the kind of analysis you are performing. If you are performing a "critical" analysis, you will import only the items that can cause a FF. An alternative is to import all equipment, including those items where a failure will not result in a FF. The logic of analyzing "non-critical" equipment is that there is a value in balancing proactive tasks with reactive tasks for items where the cost of failure is high. In other words, there are instances where an ounce of prevention is worth a pound of cure. (If the system or software being used is set up to compare the cost of PM to the total cost of failure, you can perform a "non-critical" analysis at the same time as you perform the "critical analysis.)

- The next step is to identify the "failure modes" for each item you have decided to include in the analysis. This point provides another choice. "Classical" RCM looks at all possible failure modes. "Streamlined" RCM looks only at those failure modes that have happened in the past and are likely to happen again in the future. These latter are called "Dominant Failure Modes". If the RCM analysis is being done to support safety through absolutely reliable operation (as with an aircraft, or a nuclear power plant) the former choice might be best. If the objective is to enhance reliability by developing a cost-effective maintenance program that focuses resources on areas offering the greatest risk to reliability, the latter approach is the most appropriate.

- The next step is to go through each item and characterize the current situation concerning each of the failure mode(s) and effect(s). This step will completely describe the current economics. Questions to ask for each mode include:

1. What is the Dominant Failure Mode?
2. What FF does it cause?
3. What is the Mean Time Between Failure (MTBF)?
4. What is the annual likelihood of an event?
5. How long with the FF last when this failure mode occurs?
6. What is the cost of repairs associated with this failure mode?

- You can now calculate the value of the current

annual risk. Again, here is a point at which choices are possible. Some RCM programs tend to simplify the risk calculations by assuming the annual likelihood of failure is one divided by the MTBF. This approach assumes that each year has a nearly equal likelihood of producing a failure. Other approaches use Weibull analysis to calculate the likelihood. Still other systems use the statistical results from modeling. At the end of the day, the objective is to quantify the reduction in risk associated with applying some form of prevention. For the sake of clarity, I will use the simplified approach in this discussion.

The calculation of value of current annual risk is accomplished as follows:

$$Risk = Impact \times Likelihood$$

$$Impact = (Hourly \text{ or daily value of FF} \times Days \text{ or hours out of service}) + Cost \text{ of Repairs}$$

$$Likelihood = 1 / MMBF$$

Extending the example from the above, assume that the boiler has only one boiler feed water (BFW) pump and:

- It has a history of failing every seven years.
- When it fails, it requires seven days to repair.
- When it fails, the boiler is completely down.
- When it fails without warning, it costs $25,000 to repair.

The value of the annual risk is:

Impact = ($24,000/day x 7 days) + $25,000
         = $193,000

Likelihood = 1 per 7 years = 0.1428 per year

Annual Risk = $193,000 x 0.1428
            = $25,571.

- The next step is to identify the possible forms of prevention and quantify their impact.

  One possibility is to implement some form of predictive maintenance. Predictive maintenance could take the form of vibration analysis, oil sampling, or any of a number of non-invasive tasks. A predictive task may not prevent an outage, but it might provide enough warning for the owner to be prepared. This warning might reduce the duration of the outage or reduce the cost of the event by avoiding some of the impact resulting from an unexpected failure. Another possibility is that predictive maintenance might limit both the cost and the duration of the outage by providing for a limited repair (say replacing a bearing, rather than over hauling a wreck).

  Another possibility is to implement some form of preventive maintenance. If the failure mode is wear out and can be reliably predicted, it may be possible to intervene and eliminate an unplanned event. For instance, if the failure occurs at the end of the seven-year life (rather than randomly throughout the seven years), it would be possible to proactively change out the pump at the six-year point or during the last planned outage before the predicted failure. This approach would significantly reduce the risk of unplanned failure.

Another possibility is to implement some form of physical change to the system, like installing a spare BFW pump, or replacing the current pump with one that is more reliable. Where risk is high and the need for reliability is great, redundant equipment is justified. With installed spares, simplified proactive maintenance tasks like operator observations may be sufficient to allow pumps to be switched in a timely manner and completely eliminate any significant risk of an outage caused by an unexpected failure.

For each of the three examples cited above, it would be necessary to calculate the annual cost of prevention and add that to the reduced value of annual risk resulting from the change, to understand the new total cost of ownership with prevention.

As an example, let's assume that the recommended solution involves the performance of some form of preventive maintenance. Assume that the PM being recommended is to be performed weekly and will cost $250 per instance.

The yearly cost of the recommended prevention would be:

PM = $250 x 52 = $13,000 per year

Now let's assume that we are confident, based on earlier experiences, that this form of PM will extend the life to ten years. Therefore the new MTBF is 10 and the new likelihood of failure is $1 \div 10$ or 0.10.

The total cost of ownership (TCO) with the new preventive maintenance is:

TCO = New risk + Cost of PM
    = ($193,000 x 0.1) + $13,000
    = $32,300

The TCO with PM is thus more costly than without PM, so it would <u>not</u> be cost effective to apply the recommended PM.

This kind of result in frequently encountered. In the example, one might:

- Look for other forms of PM that are less costly.
- Consider performing the task less frequently.
- See if other tasks are more effective at extending the life and reducing the likelihood of failure.

When considering the alternative of installing a spare pump, another way to compare the cost of installing the spare pump with the long-term value would be to compare the Net Present Value of all future risks that are avoided with the current cost of installing the spare pump. The saving in future risk equals the risk without the spare pump, minus the risk with it installed.

Each form of prevention may produce a different form of reliability improvement. Some forms of prevention may reduce the likelihood of failure. Some may reduce the impact of the failure by reducing the duration of the outage, or the extent of the FF. Still others might reduce the extent of the damage (to the pump) and the resulting cost of repair.

The objective of the analysis is to:

• Understand the current relationship between failure modes and effects.
• Identify some form of prevention that attacks the failure mode and/or reduces the effect.

The next step in the RCM analysis is to continue this analysis through all the items that have been chosen for analysis and imported.

Two final steps are needed to produce results that are both practical and pragmatic. The first final step is to compare all the new recommendations for the proactive maintenance program to the current program. It is frequently best to roll the new program out over a period of time so that personnel have time to absorb the changes.

The second of the final two steps is to overlay all proactive maintenance programs onto a single spreadsheet. Such a layout will frequently make all programs more efficient by scheduling some tasks a little earlier and other tasks a little later, so that crews can address more things at one time.

> There is an old joke by comedian George Burns in which he said, "You can tell you are getting old by, when you bend over to tie your shoe, you think about what other things you can do while you are down there." Efficiently organized maintenance takes advantage of the wisdom of this joke.

In organizing a RCM program, there are a variety of other choices that can be made in terms of functionality and resulting recommendations.

One of these choices is the inclusion of an additional step at the end of the analysis of each item. That step simply asks the question "Is this proposal good enough?" It is possible that despite the improvement that will result from your recommendations, the performance will still not be good enough. If the answer is that the performance will still not be adequate, it might be necessary to perform Root Cause Analysis, or some other more directed form of engineering analysis, to produce the required perform-ance.

Another choice you can make in organizing your RCM program has to do with who does the work being recom-mended and how the work will be done. Many of the proactive maintenance tasks can be performed by jour-neyman craftspersons, or they can be simplified and per-formed by the operator. An example is vibration analysis. Ask yourself: "Can my objective be equally well met by an operator simply laying his hand on the bearing housing each shift to see if the vibration level has increased?"

This approach may eliminate the need for an expensive vibration technician. It will also encourage the operator to become more intimately familiar with his equipment, and allow the task to be done once or more every shift, rather than on some less-frequent basis.

If RCM is used in this manner, it provides an excellent tool for populating the operating rounds in plants using Operator Driven Reliability or Total Productive Maintenance.

In conclusion, RCM is most often viewed as a tool to identify an optimized program of predictive and preventive maintenance. The objective is to achieve the lowest overall lifecycle cost, including the cost of asset loss resulting from failure-caused outages. RCM is designed to exploit the maximum inherent reliability of the current system

If you are smart, like George Burns, there are some other things you can do while conducting a RCM analysis:

1. Further improve inherent reliability by identifying and addressing current performance limiters that do not meet expectations.
2. Optimize the use of all the resources in your organization by identifying simplified tasks that will enhance the reliability performance of your equipment.
3 Force your overall maintenance program to make the transition from being reactive to being proactive.

Weibull Analysis

When writing this book, it was suggested that I take a stab at de-mystifying Weibull Analysis. One valid comment I heard was,"Although a lot of people use it, probably few people really understand it". Based on my experience, that statement is probably true.

Speaking for myself, I took only one course in statistics during my college career. While I received an A in the course, it could as easily have been a C or a D. All I really hoped to get out of the course was me. When it comes to statistical analysis, my mind reacts as it does to classical music. There is an on-off switch that immediately turns off, and it is a struggle to get it turned back on.

Contrary to my natural tendencies, I firmly believe that statistical analysis, and particularly Weibull analysis has an important role to play in reliability analysis and management. Weibull analysis is capable of predicting the rate and timing of future failures. More importantly, Weibull analysis enjoys the reputation of being accurate and reliable, so it is widely accepted. The fact that it is widely accepted is no small issue.

Assume, for example, that you have purchased 100 of the same kind of device (say an XB-22) or that you have purchased a large number of equipment items that contain the same device (an XB-22). Further assume that the failure of an XB-22 results in the loss of the income-producing capability so failures are unacceptable. Now assume that your XB-22s have started failing.

Now let's say that you have purchased a portion of the population of the XB-22's in each of the last five years, and that you have good records of the date when each device was placed in service and of the date that each failed.

Your supplier is saying that he received a bad batch of XB-22s last year and that the failures you are experiencing are all from that single quality spill. Further, your supplier believes that all the defective XB-22s have worked their way out of the system so there is no reason to take further action.

Let's assume that your warrantee expires after three years, and that time is fast approaching for a portion of the population.

Many questions are running through your mind:

1. Is this a limited-quality spill or will the failures continue at an unacceptable rate?
2. Should my supplier be responsible for the costs associated with this problem?
3. Has there always been an inherent defect that precludes the time limitation expressed in the warrantee?
4. Will the XB-22 make it through to the planned overhaul or will we need a planned "maintenance event" to avoid unplanned failures? If so, when?

All these are fair questions and in a typical situation the conclusion drawn by the owner is likely to be different from that of the supplier. Unless there is some basis upon which both the owner and supplier can agree, it will be impossible to assign responsibility.

The result will either be:

- The owner will absorb the cost of failures and spend the money to address the problem, then believe that the supplier acted in an irresponsible unresponsive manner.

- The owner will have enough clout to force the supplier to accept responsibility and pay for replacments. The supplier will feel cheated and will try to recover the losses on the next deal.

In either event, the relationship will suffer and future business is in jeopardy.

The answer to this dilemma is to assemble the failure history and create a Weibull plot. Both the supplier and owner should agree to interpret the plot together and abide by the results.

Rather than providing a feeble, less than adequate, description of how to construct and use Weibull Analysis, I will provide a brief description and refer the interested reader to a comprehensive reference.

A Weibull diagram is a log-log plot of the cumulative percentage of failures (y-axis) versus Life time to failure (xaxis). This plot produces several useful characteristics concerning the useful life of any device:

1. The Beta (ß) or slope of the Weibull curve indicates the class of the prevalent failure mode during each portion of the life.

   a. Beta (ß) < 1.0 indicates infant failures.
   b. Beta (ß) = 1.0 indicates random failures (not age-related)
   c. Beta (ß) > 1.0 indicates wear out failure

2. The Beta or slope can change during the life of a device, showing that as it ages it can have two (or more) classes of failures based on different failure modes.

3. The age at which 63.2% of the population has failed is the characteristic life, Eta ($\sum$).

4. If an overhaul is projected at a specific age, the intercept of the Weibull curve with that age will iden-tify the percentage of the population what will have failed by that time.

Returning to our example, create the Weibull plot as fol-lows. Assume that the total population is 100. Then each failure constitutes 1% of the population. The cumulative percentage when one has failed is 1%. The cumulative percentage when two have failed is 2%. The cumulative percentage when three have failed is 3%. And so on. Create a table that is sorted by increasing time to failure and the cumulative percentage of failure by that time.

Life to Failure (Days)	Cumulative Number of Failures	Cumulative Percentage of Population
60	1	1
180	2	2
210	3	3
250	4	4
500	5	5
550	6	6
600	7	7

Assume that our Weibull plot for the XB-22 failures looks as follows after the data is plotted on a log-log plot:

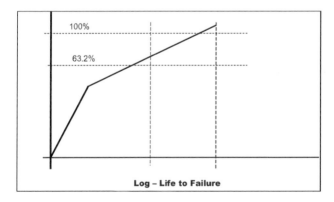

This plot tells us:

1. There are two failure classes. One is infant failure (ß< 1.0) and one is wear out (ß> 1.0).

2. The characteristic life is shorter than the warrantee period.

3. Greater than 100% of the devices will have failed before overhaul.

Based on this analysis, both the owner and the supplier should agree:

- The characteristic life is less than the warrantee period so replacement of the device should be covered under the warrantee.

- If the device is one (such as a piston or crankshaft) that is normally expected to survive from overhaul to overhaul, this device is not meeting expectations and needs to be upgraded. If not, the system will require two major maintenance events in the span of each overhaul instead of just one.

For additional information concerning Weibull analysis, the reader is referred to the New Weibull Handbook included in references at the end of this book.

Total Incident Reporting – Pareto (TIRP)

On several occasions in the past, I have been cruising along doing all the things I thought were necessary to steadily improve reliability when out of the blue I was surprised by a problem I did not know existed. In those situations, I came to understand that having all the conventional reliability processes (RCM, RCA, Lifecycle Analysis, etc.) in place does not necessarily prevent surprises associated with day-to-day problems that cost the company a lot of money. Sometimes individuals working in plants, or in tightly- knit organizations, think it would be impossible for big-ticket problems to get past their awareness. That theory may be true, but there are plenty of examples to the contrary.

I recall one situation in which short, but chronic, outages were being caused by a single equipment item. This state of affairs was viewed as a nuisance and it was thought to result from a variety of different sources. When a total incident-tracking system was installed, it was found:

1. The source of all failures had a common cause.

2. Although the unit itself was down for only a short time, the outage drove product off-spec. It took a long time after re-start to get the product back on-spec. It required even longer to re-run the off-spec material. All the time the off-spec material was being re-run, the fresh feed capacity was severely limited.

As it turned out, this "nuisance" problem was the most expensive problem confronting the plant. Until all the incidents were reported, and until all the costs associated with incident were properly attributed to the proper cause, it was impossible to identify the source of losses.

The starting point for a comprehensive Total Incident Reporting program is the determination of "best practical" performance. The best practical performance is the combination of the maximum reasonable income and the lowest reasonable cost. The combination of these two characteristics should result in the greatest possible margins.

An "incident" is anything that gets in the way of achieving the "best practical" performance, independent of its source, and if it results in an added expense, or if it prevents greatest asset utilization, or production, or both.

> **Incident:**
> **In the context of this discussion, an incident is any reliability-related event that results in added costs or in lost profit opportunity. For instance, if an event results in added maintenance costs, or in product being downgraded, it is an incident.**

In this system, each incident is recorded and all costs resulting from that incident (either added expense or lost opportunity) are recorded with the incident. It is best if the database also records some additional information like equipment involved (tagged by equipment ID number so that it can be sorted on this field), apparent cause (this field should be a small number of predetermined causes that can be selected from a table, again facilitating sorting), and other information the user will find helpful.

Over time, the database will contain a number of entries. The entries should be sorted by total cost with the highest at the top. Once sorted, a Pareto Analysis should be

> **Pareto Analysis:**
> In concept, a **Pareto Analysis** is based on the belief that 20% of any population of failures accounts for 80% of the total cost. The value results from leveraging resources. By applying limited resources to the top 20%, the vast majority of savings can be gained.

done to determine which of the items on the list should be investigated further.

If Pareto's rule holds, only 20% of the incidents will account for 80% of the total costs. Once Pareto's rule has been applied, it is likely that a number of the incidents in the top 20% (or the 80% of costs) are already being addressed by one of the other reliability programs. For the incidents in that group that are not already being addressed in some manner, a person should be assigned to investigate and take appropriate corrective action.

The real value of TIR-P comes from the secure knowledge that none of the truly costly problems is being ignored. Some of the effectiveness of a TIR-P program is associated with the way it is organized and executed. Such a program works best if there are regular (monthly) meetings to review the items falling within the top 20% (or 80% in terms of value). During those meetings, assignments are made for follow-up and progress on past months' assignments are reviewed.

Reliability Availability Maintainability Analysis (RAM)

As discussed earlier, when many people use the term "reliability", they are actually thinking of a characteristic that includes aspects of reliability, availability, and maintainability. In fact, these are three inter-related but separate characteristics. RAM analysis is a process that addresses all distinct aspects of these three characteristics.

- Reliability is a measure of the instantaneous likelihood that a system or device will fail in a specific period of time. For instance, if there is a 10% likelihood that a device will fail in a one-year period, the reliability for that period is 90%.
- Availability is the ratio of uptime or operating time to total time. If a device has a planned shutdown last ing one week in every ten weeks for maintenance, the availability would be 90%. If that same device also experiences an unexpected outage once during every operating cycle and that outage lasts one week, the availability would be 80%.
- Maintainability is a measure of the ability to restore the inherent reliability of a system in a ratable period of time. For examples, we can use responses you might get from your automobile mechanic. If you took your car to the mechanic with an unusual prob lem and he were to say, "I don't know how long it will take, but it will be reliable when I finish", the car is not maintainable. Similarly, if the mechanic were to say, "I can have it back to you in four hours, but I

don't know how long it will last", it is not maintainable. For a system to be maintainable, the mechanic needs to be able to say, "I can fix it in X hours and when I finish it will be reliable".

Although the elements are inter-related, different techniques are used to address each of the three characteristics.

For instance, RBD and RCM are good tools to improve reliability. Improving reliability is likely to improve availability because there will be fewer unplanned outages. But any availability improvements are just an added bonus. The true objective of performing RBD analysis is to provide more-reliable components and a more-reliable configuration, so that the overall system is more robust. The true objective of RCM is to identify predictive or preventive tasks that will intervene before an unexpected failure occurs. When improved reliability eliminates unplanned interruptions, availability will improve. On the other hand, there are more direct ways to improve scheduled availability.

Scheduled availability or the percent availability resulting from scheduled downtime is determined by two things: run-limiters and duration-setters. Every system has a component that determines the maximum allowable run-length between maintenance outages. This component can be called a "run-limiter". Every system also has a component that determines the minimum amount of time that will be required to perform outage maintenance. This component is the outage "duration-setter". Identifying the run-limiter and the duration-setter will allow you to modify the elements that directly control the scheduled availability. This control ensures longer runs and shorter outages, leading to higher availability.

In many instances, scheduled outages occur before they are required by the run-limiter. Some regulatory requirements force plants and equipment to be inspected or maintained at intervals far shorter than would be required by the equipment. In others, the timing of expansion projects or other enhancements drive the timing of scheduled outages. Allowing either to determine the timing of scheduled outages will leave money on the table. If regulatory requirements are driving outages, it is valuable to understand specifically which components must be inspected and if there are alternatives to removing equipment from service. Frequently, over-aggressive interpretation of regulations results in more-frequent outages than are needed. In addition, projects should be timed around scheduled outages rather than vice versa. Most companies have a number of projects competing for capital funds. The order of implementation should be based, in part, on harvesting the maximum benefit from current investments.

Once imaginary limits have been removed, it is possible to begin looking for real run-limiters. Run-limiters come in a variety of shapes and sizes. Sometimes, key equipment must be cleaned to keep it serviceable or efficient. Sometimes, wear and tear makes it necessary to inspect or maintain components that are inaccessible without an outage. Sometimes, catalyst or other contained material has a limited life and must be changed at set intervals. In any event, it is often economically possible to take steps that will modify the impact of the run-limiter. For instance, if the run-limiter suffers wear and tear, it might be possible to redesign the run-limiter to make it more robust and able to operate longer between scheduled outages.

One point to consider when spending resources to improve a run-limiter, is that there is frequently another run-limiter operating closely behind the first ... and a third closely behind the second. As a result, it is important to understand the expected life of all components that are nearly at end-of-life, when scheduled maintenance is conducted. To harvest the return on investment for improving each run-limiter, it is important to know exactly how much the interval between scheduled outages can be extended.

In addition to the run-limiter, every system also has a component that takes the greatest amount of time to inspect or repair, the duration-setter. Typical Critical Path Planning (CPM) is a good way to identify the duration-setter. When the duration of repairs to this component are strung together with the preparatory and concluding tasks, the continuous string of work creates the critical path or longest work path for the outage.

Those familiar with critical path planning of outages will recognize that there is typically one "critical path" in every schedule, and a number of paths that compete for having the critical path duration. There is almost always a way to reduce the critical path duration, but it is important to understand that the schedule will be improved only by the amount of time gained between this path and the next longest duration-setter. To further improve the schedule, it will be necessary to address that duration-setter and so on and so on. It is possible to improve the duration up to the point that the cost of the effort to improve the schedule is greater than the profit available from increased availability.

A variety of techniques is available to reduce the critical path duration:

1. It may be possible to work more hours per day or more days per week on the critical-path tasks.

2. It may be possible to work a larger crew on critical-path activities.

3. It may be possible to organize critical-path work so work continues during slack periods (breaks and lunch) on critical-path tasks.

4. It is possible to perform some of the more time-consuming aspects of critical-path work outside the outage period. (Such as pre-fabricate critical components)

5. It may be possible to redesign the facility or asset so that critical-path components are redundant and shared systems require only a short outage to switch connections.

When performing a RAM analysis, it is frequently easy to identify areas where reliability and availability improvements can be made. Opportunities to improve maintainability are not so obvious. The key elements of maintainability are: 1. Restored inherent reliability and 2. Ratable duration for repairs. When I think of maintainable systems, I most often think about changeable modules in military equipment. If a radio fails on a submarine, the technician simply pulls a module and replaces it with another module he knows is good. The inherent reliability is restored and the time to repair is known.

While none of us have as much money to throw around as the US government, we can take some steps that follow their lead. An example involves computer control systems. Frequently, systems that duplicate on-line systems are needed for training. The panels in training systems provide "hot running spares" for on-line systems. If the training facility is located close to operating facilities, common equipment can be easily exchanged when there is a suspected problem.

It is common for engineering and maintenance personnel to claim that they have completed much of the effort included in a RAM analysis, and that is often true. As with other reliability programs, the ultimate value is the result of discipline and rigor. For a RAM analysis to be effective, it is important to go through the entire system on an element-by-element basis. For each and every equipment item, it is important to identify the:

- Reliability
- Availability as a stand-alone item
- Maximum run length between outages
- Minimum time to inspect and repair during an outage

- Mean Time Between Failures
- Mean Time To Repair (non-outage)
- Capacity

The information described above should be set up in a spreadsheet that can easily be manipulated to highlight the ranking order of each element. Comparison of elements at the extreme limits will identify areas of opportunity. For instance, a simple comparison of capacity (or thru-put) will identify bottle-necks. When elements that are the cause of capacity bottle-necks are also reliability-limiters, the system has no chance for "catch up" by over-running after interruptions. The element that is out of service for the greatest amount of time is also the element least able to catch up.

Precision Maintenance

For a moment, think about the engine in your car. Think about the pistons, the cylinders, the valves, the camshaft, the crankshaft and the bearings. Think about all these things as you are traveling down the highway at 75 miles per hour at night, in a blizzard with your family in the car. Think about how precisely things need to fit together, and how much you depend on the reliability of your engine. Think about the attention and care that was given when the engine was assembled.

Now assume that you own a car that was used when you purchased it. Further assume that the engine had been rebuilt sometime, while the prior owner had it. Now think about the attention and care that the engine was given when it was being rebuilt.

Compare the mental image you had of the first assembly of a new engine with the image you had of the engine being rebuilt. What were the differences, or do you believe that both procedures were substantially the same? Were the bearings replaced? How were they fitted to the shaft? Were the pistons replaced? How were the replacement pistons chosen? Was there an effort made to ensure they were balanced? How much wear was present on the surfaces of the camshaft? Were the valves worn? What was the condition of the valve springs? For that matter what did each person performing the work look like? Did the mechanic have clean hands when assembling precision components? Is the shop clean?

There are a million questions one can ask and the answer is simply that a rebuild is unlike the initial assembly. And you only have the right to expect what has been specified.

Increasing numbers of companies are discovering the value of a process called precision maintenance. Precision maintenance is a process used in rebuilding or overhauling equipment that incorporates the following elements:

1. The dimensions of each critical tolerance, fit-up, and clearance are identified by the manufacturer. The manufacturer identifies the normal conditions and the minimum standards needed to ensure that advertised performance can be achieved.

2. During assembly, "as-left" conditions are recorded as needed to provide a basis for calculating deterioration rates in the future.
3. During disassembly, "as-found" conditions are recorded as needed to compare with the last "as-left" conditions, to determine wear and wear rates.
4. A "deterioration rate" (DR) is calculated for each critical wearing component using the equation:

$$DR = \frac{\text{"As-found" - "As-left"}}{\text{Time in Service}}$$

5. In preparing for assembly, the expected "end-of-run" (EOR) conditions are calculated using the following equation:

$$EOR = \text{Current condition} - (\text{DR x Targeted run length})$$

6. If the end-of-run conditions do not meet the minimum standards set by the manufacturer, the affected worn component is replaced or reconditioned.
7. In addition, all replacement parts are inspected for condition. For example, if pistons are replaced, they are checked for balance. If pistons are beyond an established tolerance, another group is selected from a much larger population until matching pistons are found.

Any company that has been using precision maintenance for any significant period of time has a number of horror stories concerning the condition of so-called "new" components. In one instance where an anti-friction bearing was being replaced for "rough" operation, ten "new" bearings were withdrawn from the warehouse and inspected. All except one operated as roughly as the bearing that was being replaced.

In another instance, a pump supplier changed sources for case castings. It turned out that the third-world supplier provided cases with casting defects that nearly penetrated the wall of the pump case. In yet another instance, improper metallurgy was used for parts and fasteners.

And on, and on. The bottom line is that although many manufacturers have adequate quality control and quality assurance procedures, some do not. When it comes to overhauls and rebuilds, you only get what you specify, and then only when you randomly audit to ensure specifications are being followed.

Lifecycle Analysis

If RCM and RAM analysis are done properly, the resulting recommendations are based on lifecycle analysis. In addition to using a streamlined version of lifecycle analysis as a part of those comprehensive studies, there are always occasions when a detailed, stand-alone lifecycle analysis is needed to evaluate several different alternatives.

As the name implied, lifecycle analysis is a comparison of costs over the entire usable life of a system. Depending on the type of system or device, the expected life may be thirty years or it may be as little as ten years. The starting point of the lifecycle analysis is to determine the usable life over which the comparison will be made.

There are two principles that must be accepted to apply lifecycle analysis:

1. The amortized value of future spending has current value.
2. Risk equals money.

If you exist in a culture that considers the value of current spending only, lifecycle analysis will have little value for you. Your organization needs to believe that the cost of future expenses, when reduced by the amount of intervening interest is the same as current dollars. You might be saying to yourself, "Everybody knows that!" Well, everyone doesn't act as if they knew that. Most people make choices based on first cost only. This is often true even of people who are viewed as sophisticated thinkers.

Another principle that is critical to lifecycle analysis is that "risk equals money". Many individuals are successful in temporarily avoiding the cost of risks so they believe they are home free. If you have properly evaluated the value of a risk, it is impossible to avoid the cost. If the risk analysis shows that there is a ten percent chance every year of a failure costing $1-million, at some point in every ten year period there will be a million dollar failure. Surviving one year is fine, but the failure will ultimately occur.

By applying these principles, risk-avoidance measures costing only a fraction of the value of the risk or of the actual failure can be identified.

The following example uses two alternatives and a lifecycle comparison covering five years.
Alternative 1 has a first cost of $200,000 and alternative 2 has a first cost of $250,000.

Alternative 1 requires maintenance every other year costing $20,000 per instance and alternative 2 requires maintenance every third year costing $25,000 per instance. This analysis covers only five years, so the second maintenance for alternative 2 and the third maintenance for alternative 1 are not included.

You might think these factors will bias the analysis and it is "cheating" to bias the results, but most assets have a "tax life". The tax life should be the basis for all analysis. If a major maintenance event falls outside the period of analysis, so be it. Assume the asset will be retired by then, and will no longer require maintenance.

It is assumed that the failure of either alternative will trigger an event costing $1-million in combined repair cost and lost profit. Alternative 1 has a reliability of 95%, resulting in a 5% risk of failure in any year ($1-million x 5% = $50,000.) Alternative 2 has a reliability of 97%, resulting in a 3% risk of failure in any year ($1-million x 3% = $30,000.)

Each annual total is brought back to current dollars using Net Present Value calculations based on a 10% annual interest rate.

Note that the present value of the first cost, all future costs and all future risks for alternative 2 is less costly

		Alternative 1						Alternative 2				
Year	Initial Cost	Maintenance Costs	Value of Risk of Failure	Total Annual Cost	Present Value of Annual Cost	Initial Cost	Maintenance Costs	Value of Risk of Failure	Total Annual Cost	Present Value of Annual Cost		
0	200,000.00			200,000.00	200,000.00	250,000.00			250,000.00	250,000.00		
1			50,000.00	50,000.00	45,454.55			30,000.00	30,000.00	27,272.73		
2		20,000.00	50,000.00	70,000.00	57,851.24			30,000.00	30,000.00	24,793.39		
3			50,000.00	50,000.00	37,565.74		25,000.00	30,000.00	55,000.00	41,322.31		
4		20,000.00	50,000.00	70,000.00	47,810.94			30,000.00	30,000.00	20,490.40		
5			50,000.00	50,000.00	31,046.07			30,000.00	30,000.00	18,627.64		
Five Year Total					419,728.53					382,506.47		

than the present value of alternative 1. This savings is despite the fact that the first cost of alternative 1 is 20% less expensive than alternative 2. This consideration makes Alternative 2 the best choice based on lifecycle cost.

As discussed earlier, many people have a difficult time ignoring the 20% up-front savings in favor of long term savings in maintenance costs, or value of the risk of failure. Businesses that always make "near term" choices can have superior near-term performance (though there is no guarantee because the failures may occur sooner rather than later). Businesses that make the best long-term choices have superior performance over the long haul.

Root Cause Analysis

Last, but certainly not least on the list of reliability tools, is Root Cause Analysis. It seems unlikely that an organization will be successful in improving reliability using any of the other tools or techniques, without first mastering Root Cause Analysis (RCA). The important characteristics are rigor and discipline. RCA provides an organization with the rigor needed to adhere to a structured process, and the discipline needed to gather and use appropriate and accurate data.

The importance of being able to find the true root cause cannot be overemphasized. Even large, well-established organizations, that are confident in their analytical capabilities, frequently do not have the ability to find root cause or the self-reflection needed to recognize this limitation.

Returning to the hierarchy that was described earlier, one has to understand the true Failure Mode to identify the actual Failure Mechanism. Beyond that, one needs to understand the Failure Mechanism to begin to drill down and identify the various levels of root cause.

When pursuing the root cause, the "physical cause" will always appear first and be most evident. Physical systems and components cannot speak for themselves so they are always first to be blamed. How many times have you heard, "That old thing, it always breaks down?"

No physical device ever designed itself, operated itself, inspected or maintained itself, so a human has to be involved in the failure at some point. It is important to emphasize that humans are involved in creating the causes of failure, but finding cause and finding fault are different things. Finding cause is a positive, constructive objective, and fault finding is viewed as negative and is a sure-fire way to close down all your sources of information.

Taking the root cause analysis to a deeper level, few people purposely do things that result in failures. The vast majority of people go to work every day with the intention of performing their jobs as well as they can and providing fair value to their employer. As a result, when looking for root cause, it is important to recognize there is something in the organization or business systems that allows or even promotes the human causes. This level of cause is called the systemic or latent cause.

The various latent causes include:
- Inadequate training
- Too little time
- Distractions at work
- Distractions from home being carried to work
- Competing objectives
- Misalignment in leadership
- Poor procedures
- Inadequate direction

Let's see how a common procedure can lead to defects. Frequently shop procedures are based on how things are expected to be done. For instance, in performing an overhaul, procedures may assume that a single person will perform all aspects of the overhaul. One set of eyes will make all measurements and confirm the correctness of all tolerances, fits, and clearances. Many people have found this approach will produce the best and most reliable results. One weakness of this approach however, is that it does not take into account any breaks or transitions from person to person that might occur. As a result, if the assigned individual gets sick, or if it is decided that the job will be worked on multiple shifts, it is possible that important information may be lost in the transition. A key fit up or clearance may be missed during assembly because the procedures do not require that all the data be maintained. Procedures are built for one approach and either management decision or pure happenstance has led to another approach.

In this example:
- The latent failure (inadequate procedures)
- Led to a human failure (failure to set proper clearances)
- Which led to a physical cause (poor fit up)
- Which can lead to a failure mechanism (maybe fatigue or overload)
- Which can lead to a failure mode (say overheated bearing)
- Which can lead to a malfunction report (feed pump – seized)

To eliminate the problem for all future situations, it is important to identify and eliminate the latent cause (as well as all other causes). To find the cause, you have to know the Failure Mode and Failure Mechanism and be willing to track the cause to its ultimate source.

CONCLUSION

It is awareness of unfulfilled desires which gives a nation the feeling that it has a mission and a destiny.

Eric Hoffer

When you began reading this book, it was because you were dissatisfied with the current situation. As described in the saying above, that is not a bad thing. Hidden in your desire for things to be better is a belief that they can be better. You can turn your belief into a vision and you can turn your vision into a reality.

This book has focused on "what you have a right to expect" and the tools needed to understand what that means. That approach can be viewed as being negative or limiting.

On the other hand, the same tools used to help determine "what you have a right to expect" can be used to create systems with characteristics worthy of higher expectations. It is much the same as the "glass half-full" versus the "glass half-empty" comparison. Your current system and its reliability provide an excellent starting point to achieve much greater performance. Once you know "what you have a right to expect" and why, you have an excellent starting point for making improvements. Your cup is more than half full, but there is room for improvement .... And that is an opportunity.

And an opportunity is all we really have a right to expect.

APPENDIX

1. Typical Malfunction Reporting and Defect Analysis System

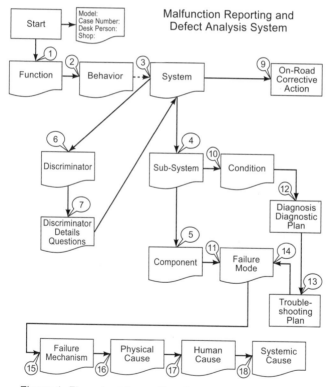

Figure 1. Flow chart for malfunction reporting and failure

	Step	Who	Development	Process
			Malfunction Reporting and Defect Analysis System Steps	
Step 1 - Malfunction Identification	0	System Engineer	Identify Equipment Number, Model, Desk Person and Shop Involved	
	0	Operator		
	1	System Engineer	Select from drop down list of Functions	
	2	System Engineer	Create list of Behaviors (or select behavior from current list)	
	3	System Engineer	Identify all systems that may produce the symptom described in behavior (or select from comprehensive list of systems.)	
	4	System Engineer	Using drawings and schematics, identify subsystems that are part of systems described above	
	5	System Engineer	Using drawings and schematics, identify the typically failing components that make up the systems described above	
	6	System Engineer	Identify subtle differences in behavior or symptoms that are needed to determine which system is involved with this malfunction.	
	7	System Engineer	Develop a set of questions that are associated with each discriminator that are helpful in pointing to one specific system or another.	
	8	System Engineer	Use the information contained in the discriminator and detailed questions to point to the appropriate system.	
	8	Help Desk		In using the system to respond to a specific failure, use the pointer described above to select the appropriate system.

Figure 2a. Malfunction reporting steps in typical malfunction reporting and defect analysis.

		Malfunction Reporting and Defect Analysis System Steps	
Step	Who	Development	Process
9	System Engineer	Based on the selected system, identify things that might be done while on the road to save the equipment failure.	
9	Help Desk		Ask the operator to apply the techniques provided to prevent a equipment failure.
10	System Engineer	For each of the Subsystems listed, create a set of typical conditions that might result in the symptom described.	
10	Help Desk		Document the most likely condition or conditions.
11	System Engineer	For each component, describe the likely failure modes (component and condition).	
12	System Engineer	Populate with possible causes and required diagnostic steps.	
12	Help Desk		Direct operator to perform steps that are needed to supplement diagnostic information and 2. Place diagnostic information in CMMS.
12	Shop Quality Leader		Access all information in this database as well as other data to finalize diagnosis of problem and create plan for troubleshooting (or repair if problem is evident).
13	System Engineer	Populate with likely troubleshooting activities and their best order of accomplishment based on experience ... revise that order based on actual results availble from this system collected over time.	
13	Shop Quality Leader		Based on diagnosis, create troubleshooting plan for shop personnel to follow.

Step 2 - Diagnosis

Figure 2b. Diagnostic steps in typical malfunction reporting and defect analysis.

			Malfunction Reporting and Defect Analysis System Steps	
	Step	Who	Development	Process
	14	Shop Quality Leader		Following the suggested troubleshooting plan, identify the actual Failure Mode and document the results with the case number.
	15	Shop Quality Leader		Collect evidence from the failed component and identify the actual Failure Mode. For Mechanical: select Corrosion, Erosion, Fatigue or Overload based on evidence.
	16	Shop Quality Leader		Based on the Failure Mode and the actual conditions where the failure occurred, identify the physical cause.
	17	Shop Quality Leader		After investigating the individuals involved and the conditions when the work was completed, identify the human cause.
	18	Shop Quality Leader		After investigating the individuals involved and the conditions when the work was completed, identify the systemic cause.

Step 3- Troubleshooting- Defect Finding- Cause Analysis

Figure 2c. Troubleshooting, Defect Finding, and Cause Analysis steps in typical malfunction reporting and defect analysis.

Malfunction Report and Defect Analysis Input Form – Part A

- What manufacturer and model does this analysis cover?

_____

1. What function does this analysis cover?

_____

2. What behavior does this analysis cover?

_____

_____

3. A defect in which systems (collection of subsystems) can produce the behavior described above?

_____

_____

_____

4. Sketch or list all the subsystems that are a part of this system:

5. For each of the subsystems described above, list each of the components that have a history of failure or of containing a defect:

* When starting an analysis of a new model and function, first review operating and maintenance manuals covering the affected systems to understand normal functioning.

** When starting an analysis of a new model and function, first review all diagnostic/troubleshooting information covering the affected systems to help complete the following sections of this database.

*** When starting an analysis of a new model and function, first review the Mechanical Problem Management Database to identify typical problems reported, that involve the affected systems.

Figure 3. Typical Input Form A.

Malfunction Report and Defect Analysis Input Form – Part B

6. Review the "behavior" and the various systems that can produce that behavior. What subtle differences in symptoms can be used to discriminate between a failure resulting from one system and a failure resulting from all others? List "Discriminators" for all system:

7. What questions can be asked of the equipment operator to help point to one system or another? List each question with the appropriate discriminator.

8. For each possible answer, identify the system that response is pointing to:

Figure 4. Typical Input Form B.

Malfunction Report and Defect Analysis Input Form – Part C

9. For each system, create a list of corrective actions that can be taken instantly to save failures or prevent outages. List the corrective actions in order of effectiveness from most likely to least likely.

10. For each subsystem, identify the typical conditions that are likely to result in a loss of function.

11. For each component listed, identify the Failure Modes (component and condition) that commonly affect the component.

Figure 5. Typical Input Form C.

Malfunction Report and Defect Analysis Input Form – Part D

12. For each Subsystem – Condition, identify steps or activities needed to produce an accurate diagnosis. This list might include reviewing downloads, equipment history, or any information available before physical contact with the equipment. For each diagnostic step discussed, identify the prognosis that will result.

13. Linked to each prognosis, there are commonly several troubleshooting steps that should be taken once physical contact is possible. List the troubleshooting steps and the order they should be taken, in sequence from most likely to least.

14. Once the defect is located, add that positive result to the database so that we can track the frequency of each failure mode.

Figure 6. Typical Input Form D.

Malfunction Report and Defect Analysis Input Form – Part E

15. Failure Mechanism – Collect evidence surrounding the failed component. Based on evidence, select the appropriate failure mechanism.

16. Based on the Failure Mechanism and the evidence, identify the Physical Cause that allowed the Failure Mechanism to exist and proceed unabated to failure.

17. Based on the evidence concerning the Physical Cause and the individuals involved in tasks related to this failure, identify specifically who should have acted differently or taken steps to prevent the physical cause. Record why he or she did not take those steps.

18. Based on evidence concerning the Human Cause and organization, and system issues associated with this failure, identify what flaw in the system allowed the human cause to exist and produce the failure.

Figure 7. Typical Input Form F

REFERENCES FOR FURTHER READING:

Abernethy, Dr. Robert B. *The New Weibull Handbook – Fifth Edition;* Robert B. Abernethy, North Palm Beach, Florida, 2004

Dailey, Kenneth W. *The FMEA Pocket Handbook;* DW Publishing Co.; 2004

Ireson, W. Grant & Coombs, Clyde F. & Moss, Richard Y. *Handbook of Reliability Engineering and Management – Second Edition;* McGraw-Hill, New York, NY, 1996

Latino, Robert J. & Latino, Kenneth C. *Root Cause Analysis – Improving Performance for Bottom Line Results,* CRC Press, New York, NY. 1999

Narayan, V. *Effective Maintenance Management;* Industrial Press, New York, NY. 2004

O'Connor, Patrick D. T. *Practical Reliability Engineering – Fourth Edition.* John Wiley & Sons, Ltd. West Sussex, England, 2002

Summerville, Nicholas. *Basic Reliability – An Introduction to Reliability Engineering* Authorhouse, Bloomington, IN; 2004

Thomas, Stephen J. *Improving Maintenance & Reliability Through Cultural Change,* Industrial Press, New York, NY. 2005

Thomsett,.Michael C. *The Little Black Book of Project Management.* AMACOM, New York, NY. 1990

Wireman, Terry. *Total Productive Maintenance.* Industrial Press, New York, NY. 2004

Wireman, Terry. *Computerized Maintenance Management Systems.* Industrial Press, New York, NY. 1994

Wireman, Terry. *Inspection and Training for TPM.* Industrial Press, New York, NY. 1992

INDEX